Becoming Nietzsche

Becoming Nietzsche

Early Reflections on Democritus, Schopenhauer, and Kant

Paul A. Swift

LEXINGTON BOOKS
Lanham • Boulder • New York • Toronto • Plymouth, UK

LEXINGTON BOOKS

A division of Rowman & Littlefield Publishers, Inc.
A wholly owned subsidiary of The Rowman & Littlefield Publishing Group, Inc.
4501 Forbes Boulevard, Suite 200
Lanham, MD 20706

Estover Road
Plymouth PL6 7PY
United Kingdom

First paperback edition 2008

British Library Cataloguing in Publication Information Available

Library of Congress Cataloging-in-Publication Data

The hardback edition of this book was previously cataloged by the Library of Congress as follows:

Swift, Paul A., 1967–
Becoming Nietzsche : early reflections on Democritus, Schopenhauer, and Kant / Paul A. Swift.
p. cm.
Includes bibliographical references and index.
1. Nietzsche, Friedrich Wilhelm, 1844–1900. 2. Democritus. 3. Schopenhauer, Arthur, 1788–1860. 4. Kant, Immanuel, 1724–1804 I. Title.

B3317.S95 2005
193—dc22 2005001171

ISBN-13: 978-0-7391-0981-6 (cloth : alk. paper)
ISBN-10: 0-7391-0981-2 (cloth : alk. paper)
ISBN-13: 978-0-7391-2320-1 (pbk. : alk. paper)
ISBN-10: 0-7391-2320-3 (pbk. : alk. paper)

Printed in the United States of America

Contents

Key for Citations:

The following abbreviations are used for citing Nietzsche's work and some of the texts of Diogenes Laertius, Kant, and Schopenhauer.

Diogenes Laertius:	
Lives of the Eminent Philosophers	(DL)
Writings by Nietzsche:	
The AntiChrist	(A)
Frühe Schriften	(BAW)
Beyond Good and Evil	(BGE) [1885]
Birth of Tragedy Out of the Spirit of Music	(BT) [1872]
Brief-wechsel	(BW)
Dawn	(D) [1881]
Ecce Homo	(EH) [1889]
Fragment of a Critique of the Schopenhauerian Philosophy	(FKSP) [1867]
The Gay Science	(GS) [1882]
The Genealogy of Morals	(GM) [1887]
Kritische Gesamtausgabe	(KGW)
Kritische Studienausgabe	(KSA)
Philosophy in the Tragic Age of the Greeks	(PTA) [1873]
Twilight of the Idols	(TI) [1889]
Truth and Lie in a Non-moral Sense	(TL) [1873]
The Will to Power	(WP)
Teleology Since Kant	(TSK) [1868]
Thus Spake Zarathustra	(Z) [1883-5]
Works by Kant:	
Anthropology from a Pragmatic Point of View	(Anth) [1798]
Critique of Pure Reason	(KRV) [1781]
Critique of Judgment	(KU) [1790]
Works by Schopenhauer:	
Paraerga and Parapelomena	(PP) [1851]
The World as Will and Representation	(WWR) [1818]

When possible, I have used Walter Kaufmann's translations of Nietzsche's work, Pluhar's translation of the *Critique of Judgment*, and Payne's translation of Schopenhauer. I have also used Daniel Brazeale's translation for some of Nietzsche's notes from the 1870's, as well as some of the translations from Gilman, Blair, and Parent. Translations cited from BAW and KSA are my own unless otherwise indicated.

Preface

During the early 1990s I began *Becoming Nietzsche* while investigating Nietzsche's philosophical reflections from his university years. At that time the 1995 reprint of the 1940 historical edition of Nietzsche's work (the Beck edition) had not yet appeared and the *Kritische Gesamtausgabe* still had not been amended to include any of Nietzsche's work prior to 1869. By chance I discovered *Teleologie Seit Kant* (1868) in the old Musarion edition of Nietzsche's work, a surprising and exciting moment for me, as I was working on a dissertation connecting Kant's *Critique of Judgment* to Nietzsche's *Birth of Tragedy*. It was then that I realized that very little investigation had been done on Nietzsche's philosophical reflections during the period of 1866 to 1868. Due to the general lack of familiarity by most scholars with Nietzsche's work prior to 1869, there was an abundance of misleading claims in the secondary literature about Nietzsche's relationship to both Kant and Schopenhauer.

In my survey of Nietzsche's surviving philosophical reflections from 1866 to 1868, his investigations of Democritus, Schopenhauer, and Kant seemed to stand out from his philosophical notebooks, as Nietzsche devoted considerable time and reflection to these thinkers. These three philosophers helped shaped Nietzsche's interests during this important formative period in his philosophical development. Written after Nietzsche's discovery of Schopenhauer but before he met Wagner or encountered Hartmann, this period of Nietzsche's surviving philosophical reflections is punctuated with the influence of Friedrich Lange, an influence that persists throughout Nietzsche's philosophic career.[1]

The investigation of this early phase of Nietzsche raises questions about what we are to make of Nietzsche's 1866-8 writings. What significance or value do they hold? Are they important for understanding Nietzsche's other works? Some of these questions are already familiar to Nietzsche specialists because of Nietzsche's other numerous posthumously published works that raise the same concerns, such as the Nietzsche lectures on language and rhetoric, *Philosophy in the Tragic Age of the Greeks*, *Truth and Lie in a Non-moral Sense*, *The Will to Power*, etc. What importance do such works have for assessing Nietzsche as a philosophical thinker, when such works were not intended by the author for publication?

Although these questions are sometimes viewed as secondary in terms of philosophic importance, almost anyone who seeks a comprehensive view of Nietzsche must address them due to the peculiar circumstances of Nietzsche's

Will to Power, a work that Nietzsche started but later abandoned. As Magnus and others have shown, Nietzsche gave up on a book to be published as *The Will to Power*, but it was still sewn together for publication after Nietzsche's death, in part for financial gain as well for political purposes, an infamous scandalous chapter in the history of ideas. *The Will to Power* provides fascinating accounts from Nietzsche's creative life, providing variations of arguments and analyses that give us a glimpse into his philosophic acumen and rhetoric. However, some scholars who deal with Nietzsche do not focus much on such work, since it never reached a final form for publication according to Nietzsche's wishes. After all, some of *The Will to Power* may be fished out of Nietzsche's wastebasket and the claim that Nietzsche was committed to a position published in such work is questionable, since he may have been trying out themes, or changed his mind during writing, etc. The works that Nietzsche published himself do not present such problems, since it seems safe to assume that Nietzsche was committed to the views he advances in the works that he willingly saw fit to share with the world.

I believe it is important to privilege Nietzsche's own published texts, as this is the work that is usually his best, but that need not imply that his other surviving work ought to be ignored, especially when he refers to some of the unpublished work in his published texts. When Nietzsche leaves notes that are contemporaneous with his published texts, if there seem to be contradictions in the positions he held, the published texts ought to win out in terms of which materials should be more seriously considered. This is not to suggest that the unpublished work is without value, but merely to suggest that the published work that is not recanted by an author usually ought to be taken as constituting the position(s) the author thought to be the most plausible. Of course one could make the case that some writers historically have opted to publish work posthumously due to consideration of personal issues (such as fear of personal safety during their lifetimes, etc.), but this clearly does not apply to Nietzsche, as his published works obviously do not shy away from controversial analyses.

Some of these same issues of interpretation that are familiar to readers of the *Will to Power* notes also apply to the current investigation of Nietzsche's work from 1866 to 1868. However, there are some differences as well. In the case of the *Will to Power* notes we also have the philosophic publications of Nietzsche that occurred during the same time periods of their composition. This is not the case with Nietzsche's work on Democritus, Schopenhauer, and Kant in 1866 to 1868, as there are no philosophical investigations that Nietzsche published during this time. Thus even if some of these early philosophical reflections are unpolished and rough, they do offer us a glimpse into how Nietzsche was developing as a philosopher during his university years. Nietzsche did have some philological publications from the 1860s, but these offer little for mapping out Nietzsche's developing philosophical acumen. Therefore Nietzsche's reflections on Democritus, Schopenhauer, and Kant from 1866 to 1868 provide us with something that is more valuable than the *Will to Power* notes in terms of understanding Nietzsche's early philosophical development and evolution, even if such work may not be as provocative or polished as some of the later *Nachlass*.

I would like to thank everyone who helped me with advice and criticisms of earlier drafts of this work, especially Richard Schacht, Wilhelm Wurzer, Rudolf Makkreel, Brent Peterson, James Porter, André Schuwer, Don Zeyl, Katherine Rudolph, Gary Scott, Ronald Polansky, Thomas Brobjer, Norbert Hedderich, and Babette Babich. Additionally, I would like to thank Dr. Wolfram Wojteki for his assistance and help with microfilm at the Nietzsche archives in Weimar.

Notes

1. Much of Nietzsche's philosophic reflection during this early phase emerges from a close reading of the first edition (1865) of Lange's *History of Materialism*. When the second edition of the *History of Materialism* was published in 1873, Nietzsche also eagerly read Lange's expansion and revisions of topics from the first edition, especially Lange's amendments on archaeology and evolutionary theory.

Introduction

How One Becomes What One Is

This text investigates some of Nietzsche's earliest philosophic reflections from the 1860s, specifically from the period of late 1866 to the end of 1868, a formative and important transitional phase in Nietzsche's intellectual development. During this phase he discovered Friedrich Lange, the neo-Kantian who probably had the most impact in Nietzsche's philosophic development.[1] However, this investigation focuses primarily on the role that Democritus, Schopenhauer, and Kant had on Nietzsche's early investigation into the meaning of life, a neglected dimension that deeply shaped Nietzsche's transition from philology to philosophy. Nietzsche investigates the significance of teleology for thinking and offers a theory of the human understanding in 1866-8, two themes that are important for evaluating Nietzsche as a philosopher. Moreover, Nietzsche advances philosophic positions on these issues that are consistent with the mature views put forth in his later published works. Most strikingly, Nietzsche's peculiar version of perspectivism already finds expression while he is still a university student at Leipzig in his early twenties.

Democritus, Schopenhauer, and Kant take different approaches to account for the existence of organic life. Democritus seems to suggest it was created by accident, denying any intelligent design present within the organism. Although Kant claims that physico-teleological judgments about the world are non-cognitive in nature, he still thinks that teleological ordering principles are needed to account for life, a claim to which Nietzsche takes exception. Schopenhauer also offers an account of organic life in which he advances a theory of the unconscious, naming the thing-in-itself as "will," an important deviation and transformation of the Kantian philosophy. Each of these thinkers provoked Nietzsche to address their respective philosophic views. While investigating and critiquing these three thinkers, Nietzsche began to formulate his own vision of how to account for the force that powers a living being.

Nietzsche tackles the issue of how to assess teleological claims and his early writings on these three thinkers share this common orientation. One of the main problems with understanding the organism is that there is an undeniable design that is present in organic life, but one that has probably been shaped by power relations and functionality, rather than a direct tinkering and shaping by an intelligent designer. Even if there is a structure present in organic life, one need not postulate nor commit oneself to any types of non-material entities. In this regard, Nietzsche's analysis seems plausible and is meaningful for dialogue with contemporary evaluations of teleological judgments.

During this early phase Nietzsche begins to form a theory of the human understanding that works in conjunction with his aesthetic perspectivism. Although the current investigation does not profess to provide an exhaustive account of every thinker who influenced Nietzsche, it does intend to shed light on this often neglected formative period of his development that is marked by philosophical analyses of Democritus, Schopenhauer, and Kant. Examination of Nietzsche's early dialogues with these primary thinkers permits a greater understanding of his philosophical development, as well as a greater understanding of his place in the Western philosophical traditions.

He was primarily steeped within a tradition of philology, but Nietzsche also enters into meaningful dialogue with the "traditional philosophers" mentioned above during the early years of 1866-8. Although Elizabeth Nietzsche is infamous for her unreliability and dishonesty as a spokesperson for her brother's philosophy, not all of her comments are misleading. She writes,

> From the beginning of the year 1866 to the autumn of 1868 we find a host of notes which prove how skeptical was his attitude towards the whole of philological study as it was then pursued; and we see how again and again he puts this question to himself in his solitude: Are the present objects of philology worth all the lives and intellectual strength which are being applied to their pursuit?[2]

Nietzsche's 1866-8 philosophic notes and projects that are left from Democritus, Schopenhauer, and Kant survive as a glimpse into his critical acumen, even if they lack the polished styles of his later published works. During this time Nietzsche's scholarship shifts its emphasis from a purely philological orientation to a broader investigation of philosophic concerns.[3] Nietzsche's writings on Democritus, Schopenhauer, and Kant survive as important evidence of his philosophic development during this phase, since these figures deeply influenced his understanding and evaluation of the history of Western philosophy. Each of these thinkers contributed to what or who Nietzsche became as a philosophic thinker.

In order to assess Nietzsche accurately as a rigorous and creative thinker it is necessary to examine his criticism of these pivotal thinkers. In spite of the apparent diversity of Democritus, Schopenhauer, and Kant, Nietzsche finds that they each left important truths and errors concerning aesthetic judgment and the meaning of organic life. These two themes, aesthetic judgment and the meaning of life, occupy an important role for Nietzsche throughout his active later period as well as within his earliest philosophic reflections.

Nietzsche's genealogy is interesting because it clarifies how original and creative Nietzsche is as a thinker. Indeed, investigating Nietzsche's origins raises questions about the limits of genealogy. Can the analysis of Nietzsche's early intellectual development reveal to us anything about where his mature views came from or how they came to be? What types of early training and materials furnished Nietzsche with the history, skills, and understanding to create his mature themes presented by Zarathustra, the will to power, the Antichrist, the eternal return, etc.? In short, how did Nietzsche become what he is?

Nietzsche writes philosophy as a type of art that presents a poetic view of the world, one that even interprets epistemology as rooted in a deeper aesthetic. If we are to take Nietzsche seriously as a creative artist, the concern of modern aesthetics to clarify the nature of creativity seems difficult to ignore. In order to shed light on the problem of Nietzsche's creativity, there seem to be some obvious starting points for inquiry: what was Nietzsche reading? Who did he talk to? What types of experiences did he have? Such questions may not fully clarify the source of Nietzsche's creative acumen, but they are starting points that help to explain some aspects of his philosophical work.

Carl Jung explicitly attacks Freud's attempt to resolve the nature of artistic creativity via psychobiographical analysis.[4] Jung maintains that no amount of psychobiographical analysis will illuminate how the visionary (a special creative artist that defies rational explanation) arrives at his creative ideas, and Jung specifically names Nietzsche's Zarathustra as an example of work by a visionary.[5] There is some literature in English scholarship that explores Nietzsche from a psychobiographical perspective, such as Karl Pletsch's *Young Nietzsche*. Pletsch provides an interesting account of the psychological events that shaped Nietzsche's hopes and dreams, but such work is not particularly useful for interpreting Nietzsche's early philosophical thinking, the primary aim of our current investigation.[6]

Nietzsche already begins to offer a theory of the human understanding in these early writings, advancing an aesthetic perspectivism and the falsification thesis that is present in his later works. He states in 1868 in his early dissertation draft on Kant that "Life is the eternal becoming, through which no light of purposive cause can shine," evidencing a deep affinity to Nietzsche's other claims about Heraclitus, a view which I believe Nietzsche never relinquished during his creative period. The view that human beings project order, identity, and purposes into the world is important for Nietzsche's theory of the understanding, and central for his evaluation of teleology. The analysis of teleological judgments proves to be an important unifying feature in Nietzsche's philosophical reflections on Democritus, Schopenhauer, and Kant during this period. It is this very feature that directs us back closer to the beginning of Nietzsche's philosophic reflections, centered in the legend of the ancient philosophical feud between Democritus and Plato that occupied Nietzsche's attention in 1866-8.

Notes

1. There is much scholarship on the importance of Lange for Nietzsche's own philosophic development. In many ways Nietzsche's philosophic reflections from the late

1860s indicate a type of neo-Kantian thought that is very similar to Lange. For a detailed investigation of Lange's lasting influence, see George Stack's *Kant, Lange, and Nietzsche: Critique of Knowledge* (New York: Routlege, 1991).

2. Elizabeth Förster-Nietzsche. *Young Nietzsche*. translated by Anthony Ludovici (London: William Heinemann, 1912).

3. This is not to suggest that there are not earlier philosophical themes upon which Nietzsche reflected and wrote. Rather, these early philosophical reflections indicate that Nietzsche's approach to the history of philosophy was infused with a new rigor during this phase, a rigor which evidences the deep influence of Lange.

4. Carl Jung. *Modern Man in Search of a Soul* (New York: Harcourt Brace, 1955).

5. Jung's analysis of Freud seems to ignore some of Freud's later reservations expressed in 1930 regarding the explanation of the nature of the creative artist. See Karl Pletsch. *Young Nietzsche* (New York: Free Press, 1991) 9.

6. Pletsch correctly claims that in the late 1860s Nietzsche realized that there were a number of decisive objections to Schopenhauer's theory of the will, but Pletsch asserts that "unfortunately [Nietzsche] did not spell them out." Pletsch. *Nietzsche*, 93. Nietzsche does in fact spell them out with a clear attack of Schopenhauer's system during this time, the primary focus of chapter two of our inquiry. In fairness to Pletsch, the nature of the psychobiographical approach may necessitate a type of focus that aims at something other than philosophical analysis. Pletsch is able to convey interesting details about who Nietzsche was as a person and how Nietzsche's interests shaped his general philosophical orientation. However, his claims about Kant and metaphysics seem to be at times oversimplified and misleading. Perhaps this occurs out of a necessity that is built into his project of conveying an intimate glimpse into Nietzsche's persona, rather than providing an analysis of the specific philosophical details that were important to Nietzsche during the early years. It is of course possible to seek out a rigorous philosophical understanding of Nietzsche's development in conjunction with an interest in psychobiography, as David Allison's *Reading the New Nietzsche* does. Although Allison is keenly attentive to the limits of psychobiography, his exploration of Zarathustra's three transformations as a personal cathartic response of Nietzsche is illuminating and convincing. Allison shows how Nietzsche's creation of Zarathustra had the important raw material of pain, specifically the anguish Nietzsche felt from his rejection by Lou Salomé. In terms of the creative process, such an impetus indicates how Nietzsche writes philosophy as a type of art (as opposed to a type of craft) in R.G. Collingwood's sense ("Principles of Art" reprinted in *Art and Its Significance: An Anthology of Aesthetic Theory* (New York: SUNY Press, 1994, 196) since a craftperson would never need pain as a vital precondition for execution, creation, or performance. Such a view lends support for the claim that Nietzsche writes philosophy as a type of art, thus suggesting a different approach to understanding his work. The creation of Zarathustra as an overcoming of pain suggests different strategies of interpretation and a rethinking of the relation of aesthetics to philosophy.

Chapter One

Teleology and the Legend of Democritus

1. Nietzsche and Teleology

All histories of Western philosophy invariably mention Socrates, a famous figure who frequently is regarded as one of the most influential thinkers in the old West. The difficulties of arriving at a definitive understanding of who Socrates was are well known, the chief of which is the apparent fact that Socrates left behind no philosophical writings.[1] Piecing together an understanding of Socrates through his disciple Plato is a complicated hermeneutic project, given the fact that Socrates appears as a literary character. The stories are there from Plato, Xenophon, Aristophanes, and Diogenes Laertius, yet it is frequently impossible to determine the exact degree to which Socrates is poetically embellished in the literature by the authors who represent him. Even if we were to believe that Plato is attempting to give us the most "authentic" version of the historical Socrates at times, it is not definitively clear when he becomes the mouthpiece of Plato, i.e., a vehicle through which Plato seeks to express his own philosophic and literary positions.

Although providing a caricature of Socrates at times, Nietzsche is also keenly aware of the conditional nature of any assertions about Socrates' character. The accounts that appear from antiquity about Socrates are based on a historical person, but Socrates only appears to us as a literary figure, formed largely in the philosophic tradition presented by Plato. Ultimately it was Plato who decided which conversations he would attempt to omit, preserve, re-enact, or fabricate in the Platonic-Socratic dialogues. This is not a modern interpretation of the relationship between Socrates and Plato, as biographers in the later ancient world already suggested many of the ideas that Socrates professes in Plato's dialogues were never even held by the historical person Socrates.[2]

What we know about Socrates' philosophy is mediated through Plato and others. Nietzsche was very attentive to the historical conditions that permitted the transmission of philosophy from the tragic age of the Greeks to the modern world. Socrates survives as a literary character, indeed a martyr for a way of doing philosophy in the ancient world. Plato's valorization of Socrates offers us an image, a representation that was molded according to the philosophic and aesthetic judgment of Plato. For all of Plato's deprecation of poetry and art in the *Republic* and *Ion*, he still invokes myths, allegories, and other poetic devices to communicate his peculiar and influential idealism. Plato is a creative artist and Socrates appears as one of the media in which he worked. Although Socrates was a historical person, Plato's philosophic imagination is largely responsible for carving out the legendary image of Socrates that survives today.

Plato's Socrates was a sage with whom many religions could identify, a mentor to the greatest idealist of all time. His student Plato left behind proofs for the immortality of the soul, as well as metaphysics that support the idea of intelligent purpose and order in the universe.[3] Later medieval traditions found such metaphysics to be friendly to ontotheology. Plato's idea of reincarnation advanced in the *Phaedo* (81c-82c), *Republic* (620a-620e), and *Timaeus* (42b) also found a compatibility with Eastern conceptions of samsara, having an affinity with both Hinduism and Buddhism.

The Socratic-Platonic view that the just person is the most happy contrasts with Nietzsche's conception of the world as tragic, for Nietzsche suggests that the world is unfair and riddled with meaningless pain and suffering. This is why Nietzsche is a tragic philosopher, in contradistinction to any religious or secular view of a moral world order. Nietzsche appears to view moral codes as originating only from the human world, a view that puts him at odds with virtually all world religions.[4]

According to Diogenes Laertius, Plato is supposed to have become a student of Socrates in 411 BCE only after Plato tossed his tragic poetry into the fire (DL III 5). Plato is also alleged to have expressed the desire to burn all of Democritus's work, a story on which Nietzsche reflects in multiple drafts.[5] Both of these stories appear in Diogenes Laertius, a sourcebook of history, anecdotes, and fiction that survives from the second or third century CE. Nietzsche focused on the sources of Diogenes Laertius to investigate the origins of Western philosophy. This early compilation of different texts furnishes the contemporary world with a strange glimpse into philosophy before the onset of the medieval world. In this pre-medieval collection, Democritus is viewed as a philosophic giant, a key point of fascination for young Nietzsche. Ultimately Nietzsche thinks that Democritus has been dwarfed, erased, and forgotten by most histories of philosophy, primarily because of his apparent rejection of religious cosmological idealism. Some truths are unpopular, especially ones that point out the fictitious nature of anthropomorphic religious teleologies. Although Democritus does not directly deny the existence of the gods, his claim that the soul is fated to decompose surely won him little support in the medieval era when virtually the only literate persons were clergy. Yet one need not even turn to the medieval era to find vast multitudes committed to a violent rejection of Democritus's position, as there are ample cases of this occurrence from the ancient, modern, and contemporary world.

Historically the promise that religion is not fiction and the soul survives the death of the body have been important claims for the survival and transmission of any philosopher's writing. Nietzsche is attuned to the reality of this political dimension in his early writings, a lesson that both Diogenes Laertius and Friedrich Lange relay to Nietzsche in the 1860s. Schopenhauer was also very much aware of this dimension, and Nietzsche focuses on Democritus as an example of this theme during his university years.

Nietzsche's Democritus investigations offer us a glimpse into his pervasive interest with the theme of teleology.[6] The nature of teleology is of primary importance for Nietzsche during this early phase, for it has profound implications for understanding human beings and other living organisms, as well as phenomenal objects, language, and cosmology. Nietzsche's inquiry demands clarification of how we are to understand ideas about order, design, reason, and purpose. The early reflections on Democritus also have important implications for Nietzsche's views about epistemology, aesthetics, and ethics which will be explored further.

There has been a tendency in the continuing history of Western Platonism to interpret Nietzsche as a somewhat radical turning point, but such a view tends to overlook the fact that there have always been philosophers who have not been amenable to institutionalized Platonism. There is in fact a constellation of thinkers lying outside the ocean of Platonism throughout the history of the West, in spite of the general widespread popularity of Plato's work in both religious and secular universities. The tendency to view Nietzsche as a radical departure from the tradition of Western philosophy is frequently accompanied by an unfamiliarity with the internal diversity that is actually present within the histories of Western thinking. Ultimately the West is not a single homogenous tradition. Some of Nietzsche's mentors from the history of philosophy have been widely ignored, leading to an over-estimated view of Nietzsche's radicality.

It is difficult to overstate Friedrich Lange's influence on Nietzsche's development in the mid- and late 1860s, for Lange had already accentuated the anti-teleological methods of Democritus as a key element in the *History of Materialism and the Critique of its Meaning for the Present*. It is Democritus who is censured by Aristotle for rejecting all teleological causes, and Nietzsche, like Bacon and Lange, was keenly attuned to the significance of such a strategy for the emergence and advancement of scientific methods.[7] Lange's masterful work is sometimes absent in major university libraries, in spite of it being one of the most provocative philosophy books written in the nineteenth century. Lange had already interpreted philosophy as a "poetry of concepts," an important philosophic move that thrusts aesthetics to the forefront.

Interpreting philosophic systems as works of the creative human imagination that bear a resemblance to art is not anything new in the modern era. Aristotle already had clearly begun to interpret Plato's theory of forms in this light as "empty words and poetic metaphors" (*Metaphysics* 99a). However, there is a further sense in which philosophic systems could be interpreted as types of creative art in the modern world, given the investigations and analyses of the knowing subject produced in modernity. After Kant, there is a focus on the active nature of the subject that suggests that human beings are creative in ways in which they are ordinarily not aware, in a much deeper and usually unacknowledged sense. *How* the knowing subject conditions its own phenomenal objects of experience is an important issue, because inquiry into this question reveals that the knowing subject is a form-giver in a very basic sense. Kant's portrait of the forms of intuition and the categories of the human understanding suggests that the unities and identities encountered within the world are complex results of a synthesizing subject, rather than anything that is simply given.

Lange's neo-Kantian view accentuates the creative aspects of the knowing subject that are presupposed in acts of representation, thus permitting one to view philosophy as "a poetry of concepts" in a deeper sense. Such a move allows one to view the discourse of philosophy as a creative achievement bearing a resemblance to artistic works of the human imagination. Indeed when Nietzsche stumbled onto Lange's work, it left such a deep impression on him that he expresses a summary of Lange's view in an enthusiastic letter to Gersdorff in late August 1866:

> Finally, Schopenhauer must be mentioned, for whom I still have every sympathy. What we possess in him was recently made quite clear to me by another work, which is excellent of its kind and very instructive: F.A. Lange's *History of Materialism and Critique of Its Meaning in the Present* (1866). Here we have an extremely enlightened Kantian and natural scientist. His conclusions are summed up in the following three propositions:
>
> 1. The sensory world is the product of our organization.
> 2. Our visible organs are like all other parts of the world of appearances, images of unknown objects.
> 3. Our real organization remains as unknown to us as the truly external objects.
>
> Thus the true essence of thing—the thing-in-itself—is not only unknown to us; the concept of it is neither more nor less than the final product of an antithesis which is determined by our organization, an antithesis of which we do not know whether it has any meaning outside our experience or not. Consequently, Lange thinks, one should give the philosophers a free hand as long as they edify us in this sense. Art is free, also in the domain of concepts. Who could refute a phrase by Beethoven, and who would find error in Raphael's *Madonna*?[8]

Lange's view was very much compatible with and influential for Nietzsche's aesthetic perspectivism. If the perception of the world and self is always filtered by a deeper process of our own organization, it follows that the objects of cognition are never the objects that are represented, thus undermining the classical Empedoclean sense of knowledge as like to like correspondence between concept and external thing. In short, we do not get the objects as they are, but only as they appear to us within the field of our own human, perspectival mediation.

By freeing philosophy from any pretensions that it offers a type of absolute truth, Lange helps furnish an alternative ground for Nietzsche to evaluate the discourse of philosophy. The question of whether philosophy itself is edifying or not resonated with Nietzsche, for such a principle for evaluation directly addresses Nietzsche's interest in whether philosophy can be in the service of life. Moreover, this type of assessment recasts the purpose and limits of philosophy by envisioning philosophy as something that is very close to a work of art.

There is good evidence that Nietzsche adopted a similar view to Lange's throughout his philosophical writings, from 1866 to 1889. Nietzsche's interpretation of the entire enterprise of philosophy as something that is distinctly aesthetic owes a great deal to Lange, a claim that will be further explicated and defended in chapter three of this inquiry. Lange proves to be pivotal in Nietzsche's philosophic reflections of the 1860s, as Nietzsche begins substantial research projects on the two philosophers thought to be most important by Lange from the ancient and modern worlds, Democritus and Kant.

To explore Nietzsche's early entrance into philosophy, three key elements will be addressed. The first consists of a brief and by no means exhaustive overview of teleology in the context of Western philosophy from Democritus to Kant. Secondly, some comments about the problematic status of the doxographic sources that furnish us with our understanding of Democritus are in order. Finally, Nietzsche's exploration of the philosophy of Democritus must be considered, with specific reference to Nietzsche's comments about Democritus's physics and ethics. This last dimension is by far the most difficult to investigate, since Nietzsche's early *Democritea* probably does not constitute a single coherent thesis. Like some of the later *Nachlass* notes that constitute *The Will to Power*, Nietzsche's reflections on Democritus do not offer a finalized work, but indicate works in progress during his transition from philology to philosophy.[9]

In recent years, there have been ongoing discussions about teleology in the philosophical community, appearing in such diverse contexts as Aristotle's physics, Kant's ethics, and more recently, philosophy of mind venues. An exhaustive account of teleology in the history of Western philosophy would require volumes, and such a project lies well beyond the margins of our current inquiry. Rather, we must confine ourselves to a simpler task, to trace Nietzsche's understanding of teleological judgment and its relation to speculative theology. "Teleology" comes from the Greek word for *telos* or end, and teleological judgment is concerned with how something acts for or is directed toward an end.

Nietzsche seems to evaluate teleological claims according to a basic distinction between "nature" and "convention," an important distinction that Nietzsche never loses sight of in his evaluation of philosophy. Such a distinction is particularly important for Nietzsche's styles of interpretation, since it is implicit in his assessment of language, as well as in his larger view about anthropomorphic perspectivism.[10] This distinction clearly has specific implications for the evaluation of ethical systems as well, but this dimension lies beyond the simpler task to be explored here, a brief overview of teleology. One of the primary concerns of Nietzsche is the status of teleological judgments: to what extent are they inventions of the human imagination?

Although there are different accounts of teleology, there is a specific understanding of teleology that emerges within the context of the Western philosophy, one that employs teleology as a proof for the existence of God. Such a proof of a director within nature is at least as old as Aristotle, although Aristotle does not appear to have any anthropomorphic understanding of a purposive force in nature.[11] However, within medieval philosophy, the argument by design is used as an attempt to prove the existence of a personal god through teleology, a proof that claims that nature exhibits design and therefore must presuppose some kind of rational designer. Aquinas employs this as one of the five ways to attempt to prove the existence of God, and a good amount of ink has been spilled writing on this topic. Among Aquinas's five proofs for the existence of God, more attention has been devoted to the design argument than all of the other arguments.

In modernity, Paley has a famous defense of the teleological proof with his example of the watch-watchmaker analogy, and Voltaire found great amusement in teleological satire, as his professor Pangloss appears as an obviously absurd attempt to reason about a supernatural designer within nature in *Candide*. Voltaire was mildly disturbed to hear the French Clergy blame the sins of the people of Lisbon for the devastation of the mammoth earthquake on All Saints Day in 1755. Concerning this issue, Will Durant writes that death, "finding its enemies in close formation, had reaped a rich harvest," since the churches were crammed on the holy day when the natural disaster struck.[12] Nietzsche also mentions Voltaire's relationship to the teleological proof in *Teleology Since Kant*, Nietzsche's 1868 natural science dissertation draft that is translated in chapter three of our inquiry.

The aftermath of the Lisbon quake seems to have been a turning point for Voltaire, since it led him to insist on a distinction between reasons and causes.[13] Given the apparent purposeless destruction of human life and property caused by the Lisbon quake and its accompanying fires and tidal waves, can one still maintain that the universe is ruled by an omnipotent and benevolent designer? Is there an intelligent creator directing everything in nature, including disasters that brutally kill children and adults indiscriminately? This is the chilling question that the natural world poses once again with the tsunami disaster of December 26, 2004: can the tragedy of the world be legitimately explained away?

Voltaire found opportunity to satire Leibniz's philosophy on this basis, a sort of comic precursor to the Kantian punchline of a blanket rejection of speculative theology.[14] Voltaire rejects this kind of ad hoc reasoning precisely because it interprets events by jumping from a physical to a metaphysical explanation of "why" something occurred. The "why" looms large here, for a meaning is presupposed to have existed, a "reason" in the sense of "purpose." Traditional folk wisdom that claims that "everything happens for a reason" specifically supports the view that the later Voltaire and Nietzsche reject, since the former view involves an equivocation with the term "reason" that vacillates between the sense of "reason" as causation in general vs. the sense of "reason" as a determined moral purpose. Nietzsche explicitly addresses this ambiguous vacillation between efficient and final causes in his work on Kant's *Critique of Teleological Judgment*, an issue to be addressed in chapter three of our inquiry.

To ask what the purpose is *behind* events of destruction refers us back to Kant's critique of physico-theology, a paradigm that is, according to the Kantian philosophy, bankrupt in terms of making any claims to cognition. Physico-theology has crept back into the fringes of microbiology at the beginning of the twenty-first century in the form of "intelligent design theorists," who claim that life cannot be explained without recourse to an intelligent designer.[15] However, there are of course first rate scientists who are also religious thinkers who have rejected the mistaken idea that an intelligent designer is a necessary postulate to account for the complexity of life.

After Kant undermined the traditional proofs for the existence of God in the *Critique of Pure Reason* (1781), the status of teleology became unclear within the critical enterprise. At the very minimum one could assert that it forever problematized speculative teleologies, insofar as such systems could purport to be scientific in nature. Yet Kant did not finish his project in the *Critique of Pure Reason*, but offers the fruition of his thoughts on teleology in the *Critique of Judgment* (1790). It is in that later work where the Kantian system is supposed to be concluded. Clearly Kant intends to complete his system by this work on reflective judgment, of which teleological judgments are a special kind.

Reflective judgments are different than determinative judgments, according to Kant, for the latter have a universal given under which the particular is subsumed. Reflective judgments indicate a more active function of the imagination, since they strive to represent a universal, a universal that is authored by an image-creating faculty. In Kant's view, the ability to come up with new hypotheses and subsume specific laws under more general laws is made possible by reflective judgment, a power that allows rational beings to organize experience and construct a framework that represents the laws of nature. It is this unifying power that captured Nietzsche's attention very early, since it suggests that the knowing subject is deeply creative in ways that are often unacknowledged. Kant's critique of the imagination is an important dimension which directly shapes the evolution of Nietzsche's understanding of philosophy in the late 1860s.

For Kant, reflective judgment proves to be central for the advancement of the arts and sciences, since it both connects judging subjects in a subjective universality within the aesthetic dimension, as well as offers human beings a basis to construct and invent new theories about the laws of the physical world. In the latter case, the judging itself does not invent the laws, but it is the condition by which we represent lawfulness to ourselves. By exploring the powers of the imagination in its creative and aesthetic dimensions, Kant's *Critique of Judgment* moves into a different terrain that has left a strange "other" legacy behind, one that has a different focus, scope, and mood than his epistemological inquiry and moral theory.

In the introduction of the *Critique of Judgment*, Kant announces that the final critique is supposed to "reconcile the seemingly divergent realms of nature and freedom" put forth in his earlier critiques. Yet the final critique problematizes the completion of the critical enterprise, and many nineteenth century thinkers recognized that there was a strange, different "Kant" in the *Critique of Judgment*. In the final analysis, the *Critique of Judgment* did not bring the critical enterprise to a neat unambiguous conclusion, but may have raised more questions than it solved. The peculiarity of this final work has spawned much inquiry into its strange status in the twentieth and twenty-first century as well, particularly in the French philosophical movements of deconstruction and post-modernism. Yet little has been done with Nietzsche's work on the very "end" of Kant's critical philosophy.[16]

Nietzsche also carefully read the *Critique of Judgment*, as his natural science dissertation draft of 1868 is written in response to developments that occurred since Kant wrote the *Critique of Teleological Judgment*. Chapter three will address this work in greater detail, but a few general remarks are still in order here to help clarify how this work coheres with Nietzsche's other early reflections. Even if it were admitted that Nietzsche exhibits contradictory ideas within his reflections on antiquity, Nietzsche can still be interpreted as offering a fairly consistent view of the human understanding throughout his philosophical and philological writings.

Kant's final critique is composed of two books, the *Critique of Aesthetic Judgment*, and the *Critique of Teleological Judgment*, two traditionally distinct topics that share some similar features. This strange unity prompts Nietzsche to offer a glimpse of his own philosophy of 1868 by concluding from Kant's peculiar construction that "Teleology, like Optimism, is only an aesthetic product."[17] Nietzsche advances claims about the dissimulative and constructive nature of the human understanding during the 1860s, a move that could be interpreted as a radicalization of Kant's emphasis on the active role of the knowing subject. The significance of the perspective of the knowing subject had been expounded by Kant even further in the *Critique of Judgmen*t, as the transcendental subject is the region "in" which the work of art takes place, within the subject's experience of a feeling that oddly lays claim to universality (KU 6). Because knowing subjects are similarly constituted, a disinterested pleasure is triggered according to Kant, one that may be shared by beings with similarly arranged cognitive faculties. This is the reason why Kant revises the critical enterprise sometime in the 1780s, as the first critique had held that all aesthetic judgments could only be empirical.[18] The later Nietzsche ridicules the Kantian aesthetic in its claim to disinterestedness, but Kant's identification of the feeling of life (*Lebensgefühl*) at the basis of aesthetic experience reveals a peculiar similarity to Nietzsche that persists in the background of this investigation.

Kant is careful to point out that teleological judgments about nature do not yield cognition, but only amount to a special type of reflective presentation. In spite of such judgments being unable to make a direct claim to knowledge, Kant still thinks that there is a compulsion to assume that there must be an intelligent cause to account for the apparent unity of organic life. In 1868, Nietzsche's *Teleology Since Kant* provides a careful analysis and rejection of this compulsion by suggesting that Kant was still unable to eliminate every theological bias from the question concerning design.

Although Kant surely did not want to portray the human understanding as an easel, Nietzsche's analysis of Kant's critique of teleological judgment indicates a radical aesthetic turn that may be viewed as an inversion and twisting free of Platonic ontology. During this early period, Nietzsche interprets the knowing subject as immersed in an aesthetic dimension that creates identities through representation. Nietzsche already begins to articulate the position that the aesthetic dimension is the most real, rather than the least real. This is why Nietzsche appears to interpret both ethics and epistemology at times as types of aesthetic enterprises (in the creative sense), a bizarre and subversive move that presupposes a deep aesthetic nature of the transcendental subject.

In the mid- to late 1860s we find Nietzsche's first outlines of a theory of the understanding, an account of the origin of reason and *how* the human understanding and reason exist in the context of nature.[19] A critique of teleology was necessary for Nietzsche to account for *how* and *why* human beings are where and what they are. What is the scientific status of teleology? Is one ever entitled to claim that there is a "reason" for something in the sense of a "purpose," if "reasons" are invented by human beings? Are they as Nietzsche infers, like aesthetic judgments, existing only within the network of the organization of a subject? As there would seem to be no aesthetic judgments without a judging perceiver, Nietzsche also implies that purposes can ultimately be relegated to a region of subjectivity, thus lacking any reality outside the context of the judging subject. Nietzsche's view is consonant with his somewhat cryptic announcement of an attempt in *The Birth of Tragedy* "to interpret the world eternally as an aesthetic phenomenon." In this regard, the attempt to draw a distinction between art and non-art ultimately refers us back to nature, where aesthetic forces are even more fundamental than the human subject, or in Schopenhauer's language, more fundamental than the individuating principle. The excess of the Dionysian and the image projection of the Apollonian that are developed later in Nietzsche's work yield a strange expansion of the domain in which aesthetics is traditionally relegated and contained.

If we may provisionally refer to the aesthetic dimension as located within the context of a frame, there is at the dawn of Nietzsche's philosophy an explosion of the frame. The aesthetic dimension for Nietzsche is neither an accessory to philosophic inquiry nor is it ontologically suspect, but serves as a condition for the possibility of identity and ontology.[20] The view that the human understanding is a kind of artwork in progress reveals a similarity of the human understanding to the easel of the artist. There is thus a sense in which teleological judgments allow us to "see" things in the world that are not really "out there," independent from the human perspective. Schopenhauer undoubtedly influenced Nietzsche's early reflections, directing suspicion toward things that are only made present by the human faculties of representation. Nietzsche postulates psychological-aesthetic categories (the Dionysian-Apollonian duality) to explain elements traditionally lying outside of the aesthetic dimension, to view even epistemology and ontology as artistic expressions, a subversive move foreign to most traditional philosophic enterprises. Rather than considering the aesthetic dimension from the perspective of science, Nietzsche attempts to consider the significance of science from the perspective of the aesthetic condition of life.

Aristotle's view of Platonic forms as poetic metaphors already had generated cause to wonder: to what extent is philosophic thinking like an artwork in progress? Although Aristotle surely postulates a kind of order that is present in nature, he does not introduce the strange ontological categories that appear in Plato's work.[21] Nietzsche's vindictive against teleology may be best interpreted as a rejection of a kind of teleological thinking that was initiated by a tradition of Platonism, one that dominated medieval theologies and found modern expression in the work of Leibniz.

Nietzsche's early work on teleology focuses on where the idea of purposiveness (*Zweckmässigkeit*) in nature originates: is there purpose or purposiveness "out there" or does it only emerge from the realm of human representation? Ultimately Nietzsche's rejection of teleology seeks to eliminate a kind of idealism that makes pronouncements about an ordered cosmos that has intentions. Nietzsche attempts to show that there is no evidence for an external intelligent designer because life itself does not emerge from any discursive intention. This view coheres with Nietzsche's larger view of the world as tragic, one that rejects any notions of karma or any other metaphysical reward and punishment systems that would be built into the cosmos. The Leibnizian view that we live in the best of all possible worlds is viewed as an optimistic fiction by Nietzsche: it is an aesthetic product, as are the metaphysical explanations that seek a cosmic design and purpose behind the destruction of Lisbon.

Nietzsche's early philology functions as a precursor to his later philosophic views about the human understanding in general. One of the most fundamental questions for the nexus of Nietzsche's philology and philosophy is—where do identities come from? To what degree are identities in general a function of the representing subject? Nietzsche's project of investigating who Democritus was thrusts this problem to the forefront. Although this question can be posed toward the individual human being, considerably more rides on this question, since Nietzsche interprets it as applying to identities in a much broader sense.

James Porter points out that in 1788 George Zoega had proposed that Homer was a symbolic fiction who did not correspond to any single person in antiquity, a claim that has also been made about Lao Tzu in the East.[22] By the early nineteenth century Friedrich August Wolf pursued "the Homeric question" in some detail, a style of inquiry that fascinated Nietzsche: was Homer merely an oral tradition constructed into a text by later editors?[23] In the 1860s Nietzsche redirects this question to Democritus: in an unreliable and uncertain history of philosophy, from where does the unified view of Democritus come? What are the preconditions for arriving at the identity known as Democritus? How have the many hands of later transcribers contributed to the image of Democritus that survives today?

The profound antipathy against Plato's philosophy that endures through Nietzsche's philosophy of the 1870s and 1880s has been well examined by Martin Heidegger and others, yet little has been done to address Nietzsche's focus on the hostility and conflict between the worldviews of Democritus and Plato in the 1860s.[24] This is an especially interesting time in Nietzsche's development, for the conflict between "this-worldly philosophy" and "other-worldly philosophy" captured Nietzsche's imagination during this early phase. Exploration into Nietzsche's reflections on Democritus reveals that the antipathy to Plato can be traced even further back into his philological writings, where he already began to formulate a theory of revenge as an underlying motive that drives Socratic optimism.

Democritus's interpretation of teleological causes had an impact on Nietzsche's entrance into philosophy, since Democritus appears to have rejected the idea of an order bestowed upon nature by an external intelligent designer. However, Democritus certainly would not argue that there are not other purposes invented by human beings. He appears only to have ruled out a certain type of discursive supernatural teleological thinking. There is an "end" for human action according to Democritus, and the worldview that has been passed down by Diogenes Laertius offered Nietzsche a kindred spirit:

> [According to Democritus] The end (*telos*) of action is tranquility, which is not identical with pleasure, as some by a false interpretation have understood, but a state in which the soul continues calm and strong, un-disturbed by any fear or superstition, or any other emotion. (DL IX 45)

Although Democritus does not appear to have had a meaningful impact on Nietzsche based on his posthumously published work *Philosophy in the Tragic Age of the Greeks*, Nietzsche clearly thought about the physics and the ethics of Democritus at length in 1866-8.[25] As Aristotle remarked that Anaxagoras was the first of the Greek physicists to view the world through sober eyes with his assumption of *Nous*, or an intelligence within nature, Nietzsche responds at one point in his notes that one could make the same claim about Democritus in his antipathy to any such claim: Democritus was a sober physicist who was not drunk from the hope of *Nous* to serve as the basis of an anthropomorphic, natural teleology.[26] How much reason or intelligence there exists in nature independent from human consciousness proves to be the central issue that is the focus of Nietzsche's early inquiry. Much hinges on this, for the answer to such a question proves to be crucial in the emergence of nineteenth and twentieth century psychology, and also thrusts the important philosophical distinction of that which exists by convention vs. that which exists by nature to the forefront.

2. The Problem of Diogenes Laertius and Democritus

For a critical examination of Nietzsche's early Democritus investigations, the text of Diogenes Laertius needs to be addressed. Nietzsche published several philological essays during the late 1860s, including an investigation of the sources of Diogenes Laertius.[27] The publication of Nietzsche's work on Theognis and Diogenes Laertius in *Rheinisches Museum* (Ritschl's journal) was very important for Nietzsche professionally, since it was largely on this basis that he was offered an assistant professorship of classical philology in Basel in early 1869.[28]

The text of Diogenes Laertius is astonishing because of its problematic status. It provided Nietzsche with a vast compilation of anecdotes, doctrines, and historical details about the philosophers in the ancient world that clearly persisted as a source for his psychological critique of Socrates in *The Birth of Tragedy*.[29] However much Diogenes attempts to present a history of the eminent philosophers, his compilation clearly is unreliable and even outrageous in some sections of the text. At which junctions Diogenes lapses into fiction proves to be exceedingly difficult to ascertain at times, and often we do not have a determinate ground for knowing precisely where a transition from history to fiction occurs, if or when it does occur. This hermeneutic problem is crucial for Nietzsche's philological inquiry in the 1860s: what may we believe from the compilation? Volumes could be written on this one question, but we must confine ourselves to a brief overview of the status of the problems here. This early philological inquiry has important philosophical implications, for it indicates that the "history of philosophy" is by no means an unambiguous phrase, a problem of which Nietzsche was keenly aware.[30] To assess Nietzsche's methods in his critique of the history of philosophy adequately, due attention must be given to this problem, the problem of the "legend" present in Diogenes Laertius.

Little is known of the life of Diogenes and the dates of his life are uncertain. He is reckoned to be a late second or early third century CE. biographer, but this date is largely calculated on his coming after Sextus Empiricus, whose dates are uncertain as well. Diogenes Laertius attempts to give an account of the history of philosophy by passing down the eminent philosophers' doctrines, anecdotes, birthdates, and circumstances of death. He refers to some of his own work that he occasionally quotes and some have argued that he is an Epicurean philosopher, yet little is known about him. His disproportionate attention given to Epicurus in relation to the other philosophers may support this point, and in this regard he is a crucially valuable ancient sourcebook for Epicurean philosophy.[31] Nietzsche himself may have been attracted to this feature in Diogenes Laertius, for some of the principle tenets of Epicurean philosophy come curiously close to Nietzsche's philosophic moods in their attempts to rule out supernaturalism, a point that will be investigated at length in further sections of our inquiry. It must not be forgotten that Democritus exerted a mammoth impact on the schools of Epicureanism, and it is hard to overstate this, especially when one considers his stress on the value of pleasure and his contributions to atomistic materialism. During this phase, Nietzsche's notes show that he was developing a nuanced view of the differences and similarities that are present in the atomistic materialisms of Democritus and Epicurus.

Diels probably coined the term "doxography" in order to refer to fragmentary secondhand transmissions of texts where the original texts do not survive themselves, so some of what we infer of the ancient philosophers is based upon reports of what the original thinker held. What is crucial to recognize for inquiry into the early history of Western philosophy is that the reliability of doxographic texts varies considerably. In other words, the texts classified under the rubric of "doxography" are not homogenous in their nature in regards to their authenticity and credibility. Attempts to understand the earliest history of Western philosophy must deal with this problem, for we have a literary landscape that contains many reports and rumors about the origins of philosophy in the old West.[32]

Within the text of Diogenes Laertius alone, there are many sections which are unreliable for a host of different reasons. The most obvious cases would be those where there are internal contradictions within the text itself about specific facts. Some of the contradictions can be explained away by the way in which Diogenes probably worked, for his specific references to hundreds of different works must have been unmanageable at times.[33] It must be remembered that Diogenes offers a compilation from many different resources, and he does exhibit a kind of critical rigor when he notes varying accounts of the facts, as well as discrepancies which are inherent in the tradition as it has been passed on to him. Yet the obvious inconsistencies at times have led many recent scholars to ignore him as a source in general.[34] As Mejer points out, earlier generations of philosophers have taken the work of Diogenes Laertius more seriously. Virtually all major Western philosophers prior to the twentieth century had some working familiarity with the work of Diogenes Laertius.

Beyond the cases of the internally inconsistent accounts that exist at various junctures in the text, there is also the problem of accounts that vary from other received traditions in antiquity that are more reliable. We do have the texts of Plato, and clearly these must be given priority over the secondary accounts of his doctrines given by Diogenes Laertius. However, there are anecdotes attributed to Plato, and comments about Plato's philosophy in Diogenes that Nietzsche found instructive and worthy of consideration. At the very minimum we can find out about received views of the early Greek philosophers in the later Greek world, even if it is the case that the later received views are occasionally off the mark. It is possible that the antagonism directed toward Plato by rival thinkers in the ancient world may not be fairly or adequately represented by Plato himself, a matter worthy of consideration, especially in regards to the Cynic reaction to Plato that has been attributed to Diogenes "the Dog" of Sinope and others.[35]

For a general reconstruction of the earliest history of Western philosophy, examining the surviving texts must be evaluated as crucial for an accurate representation when this is in fact possible. However, for almost all the pre-

Platonic philosophers this is not possible. When it is the case that Diogenes' accounts of the earlier philosophers agree with other ancient sources, we may provisionally reconstruct the portraits of the earlier thinkers, but of course these can never attain the degree of hermeneutic certainty that is afforded by having primary sources. A third circumstance that we must address are those doctrines, anecdotes, and dates in which Diogenes is our only source: in some cases the problem of what we may believe proves to be most interesting, for there are junctures where we simply do not know whether they can be believed, having neither anything else with which we may compare them nor any means of validating their authenticity. This last case of texts is a special hermeneutically ambiguous region, for there is insufficient evidence to conclude whether such textuality is history or fiction. For the current investigation, this ambiguous region needs to be addressed, since Nietzsche's attention turned to it frequently, especially the stories present within Diogenes Laertius that report the motives of the agents that burned texts in the ancient world.

When we have Diogenes Laertius as our only source, there are some cases that may be referred to as "legend" in a traditional sense of the term, stories that have been passed down from generation to generation of which no one knows whether they are historically accurate. The ground for evaluating the authenticity of some aspects of the "legend" simply does not exist. It would be much easier to address the meaning of the phrase "history of philosophy" were it the case that no legends (stories about the philosophers transmitted secondhand from the earliest histories of philosophy which may or may not be true) existed, but this simply is not the case. There is a large collection of accounts that would fall under this rubric of "legend" and they problematize the phrase "history of philosophy," by making it scarcely possible to make determinate claims about some of the origins and authenticity of the discourse that has been passed down to us.

The reflections of Diogenes Laertius offer an alternative view of the history of the old West, for beyond the concern with what arguments or positions a philosopher held, he clearly is concerned with "who" the philosopher was, something that could only be captured in assessing the life activities of the philosopher, for the decisions and actions of the philosopher may reveal something significant about the particular character of a thinker. The "who" proves to be important for the inquiry of Diogenes Laertius, and clearly this approach had an influence on Nietzsche, especially in Nietzsche's inquiry into the personalities of the pre-Platonics in his work *Philosophy in the Tragic Age of the Greeks*. In *that* early work, Nietzsche wanted to know what kind of individuals the philosophers were, something that could only be determined by their life activities, what they did or said in real, unrepeatable circumstances. This kind of inquiry could be thought of as "existential," for it deals with the unique unfolding of "who" the individual philosopher is. Although it has little to do with "existentialism" as a systematic phenomenological ontology, such a method may be genuinely characterized as "existential" insofar as it focuses on the individual's character. Nietzsche later remarks that the personalities or character typologies of the ancient Greek philosophers are something that is irrefutable. Such aspects need not be thought of as mere trivia, but may be viewed as important for understanding whether there is a personal relation between a philosopher and a philosopher's system.

Diogenes Laertius offers much about the existential characters of the earliest philosophers, something that Nietzsche also adopted as a strategy for investigating what the person had done or achieved in the course of life. It is easy to dismiss this question of character as a non-rigorous psychobiographical approach for evaluation of a philosopher, yet such a method of inquiry need not lapse into arbitrary ad hominem arguments, nor be fueled by a spurious dogmatism or voyeurism. Such concerns are relevant for Nietzsche's unpopular question: Is the philosopher's worldview a psychological coping mechanism?

The *Lives of the Eminent Philosophers* attempts to reveal something about the personalities of the ancient philosophers by anecdotes. From one perspective this might not seem like it could contribute anything toward a rigorous account of the history of philosophy.[36] However, eliminating this kind of interpretive writing as an acceptable way to investigate philosophers risks the danger of excluding something that may be genuinely philosophical, as well as psychologically interesting. Daniel Brazeale points out that Nietzsche's untimely meditation, *Schopenhauer as Educator*, in spite of being a book about the philosopher for whom it is named, refers to almost none of the specific technical details of Schopenhauer's philosophical system.[37] Yet there still exists a kind of inquiry that asks *who one is* that is a viable ground for evaluating the thinker as a person. The larger issue of the connection between what a thinker thinks and who a thinker is need not be rejected as a spurious psychologistic concern.[38]

Nietzsche valorizes the "powerful masculine seriousness"[39] present in Schopenhauer, and such a characterization warrants further evaluation. What does it mean to be a "real man" in this evaluative sense?[40] Nietzsche suggests that independent personalities characterized by strength may indicate a type of health. In some respects this is similar to Aristotle's account of virtue as a character trait that provides one with the most lasting and durable of all human goods, a type of self-autonomy. Such a trait suggests the ability to stand alone for the pursuit of truth without a psychological need for validation from other philosophers or the general public. This autonomous sense of self-sufficiency is important for both of these thinkers, for it takes courage and power to sustain oneself by affirming unpopular and ugly truths, something Nietzsche seems to associate with the masculine.

Nietzsche views Schopenhauer as exhibiting a kind of integrity by holding fast to a type of serious thinking that has no nostalgia for empty ideals. This marks a type of intellectual integrity that may not always be amenable to the politics of academic philosophy. My exploration of this dimension does not condone Nietzsche's sexist comments nor endorse the patriarchal exclusion of women in the history of philosophy, but merely attempts to highlight why Nietzsche refers to a "powerful masculine seriousness" as a virtue. There is a sense of a powerful independent character that Nietzsche finds in both

Democritus and Schopenhauer, one that raises the question: how can the sage exist outside of the academy? Neither Democritus nor Schopenhauer needed to sell themselves or their philosophical reflections for the sake of a livelihood. Neither of these thinkers felt compelled to develop allegiances with organized religions for moral or financial support. Schopenhauer at times even seems to take a perverse delight in making fun of the absurdities of organized religion. Although this may not seem extraordinary in the contemporary world, prior to the twentieth century there were comparatively few philosophers who were mercifully free from the political influence of organized religion.

Nietzsche is interested in how philosophical thought has been conditioned by power structures. In many senses, this first part of our inquiry is a book about religion, politics, and censorship. As Nietzsche points out, there are many different ways for books to be destroyed. Book burning and fanaticism go to frightening lengths. Diogenes Laertius's biography of Protagoras provided a lesson to Nietzsche about power structures that permit or bar the evaluation of philosophic texts.[41] Burning texts because they are perceived to be dangerous seems absurd, but these stories appear at multiple junctures in Diogenes Laertius, and Nietzsche was specifically fascinated with Plato's connection to book burning.

Whether philosophy may be a sign of sickness rather than greatness, a kind of refuge for the uninteresting personality or a form of escapism, remains a profoundly unpopular question about academic philosophers that Nietzsche's legacy has left behind. From this perspective, philosophic questioning may be a sign of decline, weariness, or escapism; it may also indicate a kind of affirmation of life. From one perspective, the "who" may appear philosophically shallow or insignificant, but the neglect of this dimension entirely could indicate that philosophy has become sterile, barren, and ceases to care about the meaning of "greatness." If philosophy scholars produce a biased account or a total neglect of this question, Nietzsche may be relevant to contemporary discussions in the sense that the value of such inquiry is not even considered a legitimate philosophical question. To adequately articulate the seriousness of this question, one would need to distinguish between the philosopher and scholar, and there are many scholars who have dominated the discourse of academic philosophy who would preclude the possibility of this question being formulated seriously. Such an undertaking would demand a radical inquiry into what *seriousness* means in a philosophic sense.

Nietzsche envisioned a new type of philosopher for the future as a cataclysmic event, and such expectations are invariably viewed with suspicion within the halls of academia. The distinction between scholar and philosopher is not thought of as significant by many thinkers today. Like Schopenhauer, Nietzsche experienced deep reservations about both philology and philosophy within the academy, and the philosophic life, style, and rigor of Democritus offered Nietzsche an alternative to the history of Platonism. Democritus's claim that "undisturbed wisdom is worth everything" struck a chord with Nietzsche in the mid-1860s:

> One becomes more cautious in speaking pronouncements of the priority of Leucippus to Democritus. It certainly is a great contribution, to conceive of a totally new intuition, but there is a greater spark to be hit upon from all the pages of Democritus. The wisdom of quiet thought, which remains undisclosed to the university student, has in the history of the sciences received little claim for its worthiness. (BAW 3 328)

Like Nietzsche's early attraction to Schopenhauer's character, Democritus also was an ideological mentor who philosophized outside of the academy. From the text of Diogenes Laertius, we have a flattering account of Democritus, as he is estimated in the highest regard among the ancients, a somewhat different assessment compared to other surviving Platonic accounts of Democritus in the later ancient and medieval worlds. The evaluation of Democritus from the medieval world becomes quite the opposite in tone, for (among other reasons) the materialism of Democritus appears in stark contrast to the theological-political agenda that dominated the idealism of medieval ontotheology. Beyond his rejection of an intelligent designer within nature, Democritus's "this-worldly" physics suggest that the soul was fated to decompose, just as the body does. As Nietzsche and Lange argue, the methods of atomism were crucial for progress in the history of the sciences, since atomism eliminates the cryptic mythological explanations of origins that Plato advances in the *Timaeus*. If one were to compare the views of Plato with Democritus in terms of their historical contribution to the understanding of organic life, the Democritean physics prove to be much closer to the truth, in spite of the unpopularity of the Democritean view from the religious perspective.

Historically theologians have policed philosophy. Although this is most obvious in the medieval world, the ancient, modern, and contemporary worlds have had their share of persons who have been denied opportunities for free expression due to their rejection of organized religious practices. The fact that philosophical reflection has been conditioned by other power relations should surprise no one, given the bloodbath of historical violence predicated on religious conflicts that persists even today. Like Diogenes Laertius, Schopenhauer observed that the conditions for the proliferation of philosophy are often bound to their compatibility with theology. Nietzsche became attuned to this dimension quite early, which one may correctly refer to as a "thought police" without conjuring up unfounded images of paranoia. Indeed both Lange and Nietzsche suggest that the dominant history of Western Platonism has been written according to an implicit value structure that sanctions and favors the very tendencies toward supernaturalism and teleology which Democritus rejects.

Are we entitled to expect the discourse of philosophy to be free from political conditions when most philosophy is done within academia? In such a context what may one expect from philosophy as a means to search for truth? What truth or truths can one get from such a medium? These are important questions for Nietzsche, since the significance of genuine philosophers is put on the scales for measure. If we were to understand genuine philosophers as rare examples of self-sufficiency, health, and greatness who radically question traditional authorities, then such persons are natural enemies of rulers who seek only to control citizens and make them obey. As Nietzsche remarks in the 1870s, if a tyrant wants to rule over a people, then the correct number of genuine philosophers to have in a state is zero. Of course there is no shortage of philosophy scholars willing to utter uncontroversial truths and form allegiances with political and religious authorities which sanction and underwrite their philosophical missions. However, Nietzsche suggests that philosophy scholars often have little to do with the genuine philosopher, and Democritus seemed to offer a model of what was possible for philosophic greatness.

Nietzsche shifts his purely philological investigations toward a new philosophical analysis of the question: *who* was Democritus? As Nietzsche would later remark in *Philosophy in the Tragic Age of the Greeks*, we measure the ancients by false standards that favor Plato and Aristotle, for they have survived by the accident that they had disciples in schools and never lacked copyists in the ancient world. What do we know of Democritus, when his work survives in the constellation of fragments by numerous ancient authors? It is alleged that Democritus was the most prolific writer of the pre-Platonic philosophers and the most widely-traveled.[42] Nietzsche notes that Democritus's developments in atomism clearly influenced Epicurus, as well as the history of skepticism, with first rate skeptics such as Pyrrho expressing enduring interest in his philosophy.[43]

The question "What may we really believe about the legend of Democritus?" is important for understanding how Nietzsche developed as a philosopher in the 1860s. Friedrich Lange clearly helped stimulate Nietzsche's interest in Democritus. In all likelihood the most important figure for Nietzsche's assimilation of the history of Western philosophy, Lange's neo-Kantian synthesis provides Nietzsche with an important critique of the human understanding, as well as a focus on the scientific and political significance of teleology. Indeed, much of the surviving philosophic reflections from late 1866 through 1868 are generated from Nietzsche's close reading of Friedrich Lange's *History of Materialism*.

Although some scholars have looked for the connection between the Democritean physics and ethics and have found little for a unifying theme, Nietzsche and Gregory Vlastos have both attempted to show a unity exists in the worldview of the fragments.[44] Democritus's controversial claim that the soul is fated to decompose, along with his rejection of an intelligent designer within nature, are not isolated, disconnected fragments of his philosophy, but may be interpreted as important interrelated parts of his overall worldview. Principles for human conduct are formulated from the fragments which do not seek refuge in any supernatural grounding for ethics.

> Parts of the ethical fragments have a free worldly-wise tone and a beautiful form. They stink neither of Stoicism nor of a high-flying Platonism, but are like Aristotle and his *metriopathia*.[45] The ethical fragments of Democritus are not

unworthy. (BAW 3 333)

Nietzsche suggests that the Democritean worldview is rooted in an ethical way of viewing things, a type of reverse fanaticism that bears a similarity to the feeling of balance in the ethics of Aristotle. Nietzsche's comment that the core of Democritus's philosophy lies in his ethical view suggests that the methods found in the Democritean system originate from the service to Democritus's vision of the good life: to be free from the disturbance of superstition. In this sense, there is an intimate relationship between the system and the thinker. Although Socrates is often credited as a turning point in the history of philosophy in his focus on the ethical dimension, the surviving fragments of Democritus indicate a deep concern with ethics as well.[46] Nietzsche observes that Democritus offered a poetic view of the world, a model of great health in the ancient world.

Like Nietzsche, Democritus offered aphorisms in a striking style, as well as a deep skepticism that influenced later Pyrrhonism. Karl Marx investigated Democritus with a substantial research project in his early twenties as well. Marx specifically interprets Democritus as a skeptic riddled with internal contradictions, from which he opts to favor Epicurus's dogmatism. Marx appears to be concerned with a philosophy of action and interprets skepticism as a type of self-refuting paradigm.[47] It is somewhat odd to find a thinker who deliberately argues in favor of dogmatism such as Marx, but he seems to have an overriding concern with making a practical difference in the world. Even if dogmatism is theoretically weak, it can be viewed as pragmatically important as an agent of change.

Nietzsche's early encounters with Democritus view the atomist much more favorably. Although Nietzsche recognizes the currents of skepticism running through Democritus, Nietzsche does not regard the Democritean worldview as incoherent, as Marx had. Democritus is a Promethean figure for Nietzsche in his courage to rule out supernaturalism. The power of reason and the drive to live an undisturbed, scientific life seem to speak directly to the young Nietzsche, across the eons that divided these two thinkers. Indeed one can find a type of joyful wisdom present in Democritus, the only philosopher from the ancient Western world who had a reputation for laughter, precisely what Nietzsche's Zarathustra proclaims as most holy.[48] Like Plato, Democritus suggests the soul is supremely valuable, but the atomist exhibits no fanatical desire to escape physical embodiment. Democritus offers a type of ethics, but one that seeks no grounding in other worlds, exhibiting at times a peculiar similarity to Zarathustra's teachings.

Although Lucretius's *On the Nature of the Universe* is widely recognized as both a poetic and atomistic view of the universe, few thinkers have examined the poetic vision of Democritus that occupies Nietzsche's attention in the mid-1860s. Later, in the closure of his May 1869 inaugural address in Basel, Nietzsche would invoke the poetry of the atomistic thesis by suggesting that "we are merely atoms."[49] Paradoxically, the poetry lies in the explanatory thesis that we are just dust in the wind, composed of small invisible particles, ultimately created out of a random swerve. Atomism is a poetry of the sublime, a recognition of the chance origin of life and its precarious unity within nature. The ability to unify diverse phenomena by "seeing" the one in the many postulates invisible uncuttables as the ground for the world we experience.

After Kant had viewed speculative metaphysics (claims about God, freedom, and immortality) as beyond the possible bounds of sense and cognition, it is not difficult to understand how one might infer from Kant's conclusions that any speculative metaphysics could be interpreted as a kind of unintentional artistic activity. Lange had begun to interpret philosophy as a poetry of concepts and Nietzsche was responsive to this neo-Kantian synthesis. Nietzsche begins to interpret both the idealistic and materialistic traditions of philosophy as types of creative enterprises. An affinity exists with Nietzsche's later philosophic concerns in terms of how one is to evaluate a thinker from the past. Can the philosopher speak across eons of time to offer something of value to the present world? This is especially significant for Nietzsche since it is during this phase that he moves beyond a purely philological interest in ancient literature, as such a historical approach risked the danger of culminating in a type of nostalgic trivia. Nietzsche begins to approach the ancients with a desire for something that could be used in the service of life in the present, a theme that would later be articulated in detail in his second untimely meditation, *The Use and Abuse of History*. Like Descartes's view that texts provide us with an opportunity to converse with the greatest minds throughout history, Nietzsche looks to literary studies with a purpose, to find exemplary models of what is possible for a great human being. Thus some parts of the past are worth forgetting, but there also are important parts that are worth remembering for the purpose of edification to meaningfully impact the present.

Nietzsche's early philosophical critiques (1866-8) have implications for the status of aesthetic judgment: in Democritus it is valorized as poetic style. In Schopenhauer, it is attacked for being employed by Schopenhauer while forgetting that it is a fictitious way of presenting things. In *Teleology Since Kant*, Nietzsche explores reflective judgment and its capacity to invent unity, identity, atoms, and purposive causes. Although Nietzsche considers different aspects of aesthetic judgment in his early reflections on Democritus, Schopenhauer, and Kant, there is a common theme present that attempts to clarify the form-giving nature of the human understanding.

Nietzsche valorizes the character of Democritus and also focuses on the poetic quality of the ethical fragments in Democritus. Beyond Democritus's aphoristic style, Nietzsche admires him for his freedom from any kind of fanaticism or zealotry. Nietzsche remarks that "Democritus was the first to strictly exclude the mythical" (BAW 4 63) in his investigation of the world. In Nietzsche's attempts to understand philosophy in the tragic age of the Greeks, we will see after this short glimpse into Nietzsche's Democritus that this proves to be of importance, for Democritus appears, at least according to Nietzsche's interpretation, as an important alternative to Platonic idealism.

3. A Glimpse into Nietzsche's Democritus (Short Translation)

Democritus demonstrated a strict scientific view and method. We commonly see a neglect of the writings of Democritus today, and with a reason—namely, since people today have not learned anything and have barren souls.[50] In and for itself there lies great poetry in atomism. An eternal shower of variant bodies appear in diversity. Movement occurs and they are woven together, so a vortex originates (BAW 3 332). . . .

He would quickly be finished with the construction of the world and ethics: the deeper problems concealed themselves from him. His will was the power that drove his sense of observation; he wanted to be finished for once and have attained the final knowledge. On that he believed then; and this gave him a self-sufficient trustworthiness. He still did not recoil in horror in observing from earlier systems an endless abundance of different standpoints. He separated from the lesser forerunners what came to him as homogenous, opting for the intelligent, the simple, and mercilessly condemned each mixing in of a mythical world. Consequently he is a confident rationalist; he believed in the redeeming effect of his system and saw the bad and the unfinished as lying outside of the system.

With that he became the first of the Greeks with a scientific character, who endeavored to explain the unity behind the abundance of appearances, without bringing in at difficult moments a *deus ex machina*. This new type had impressed the Greeks. As one of a kind he was devotee to science, having worked as a restless nomad full of self-denial and finally into an incredibly old age, he was the harmonious improvement and contrary to the medium. Democritus himself felt this as a new principle by which to live; he valued a scientific exploration higher than all the riches of Persia.[51] He believed he had found the scientific life to be the goal of all eudamonism. From this standpoint he threw out the life of the masses and all earlier philosophers. He derived the pain and sadness of human beings from the unscientific life, all from the fear of gods. From this Empedocles and his dark mythology must be thought of as the great forerunner of Democritus. He even had an unconditional trust in the power of disclosure [*Schlußkraft*] of the *ratio*; the world and human beings were to him, as he believed, unveiled, and therefore he threw out the coverings and the borders, and posited precisely the same *ratio*. A scientific life was at that time a paradox. And Democritus declared himself as an enthusiastic apostle of the new teaching. Thus the poetic swing of his presentation, which we can easily observe as striking. The poetry does not lie in his system, but in the faith which fastens it all together.

The enthusiasm is to be explained as a similar kind as that of the Pythagoreans for number. The first beginnings of scientific knowledge by the Greeks would become considered through drunken eyes.

Thus Democritus maintained an ethical worth in his way of viewing things; he believed in the happiness of human beings when they would walk through life with his scientific method: from this, one is reminded of August Comte. This belief made him into a poet, and the poetic moments are rarely included in the teaching itself. With his scientific method he dedicated his whole life only to the struggle to fathom all types of things. Thus he was also the first to elaborate systematically the subject matter of all of the sciences.

Democritus was a beautiful Greek nature, like a statue appearing cold, but fully concealing his warmth . . .

The ethical writings thus show how the kernel of his philosophy lies in the ethical side. His ideal is unencumbered (*Ungehudeltsein*), a quiet scientific life. He reflected closely upon the different aspects of life from which people would become disturbed (BAW 3 348-50).

The mythical disturbance. Rationalism.

The soulful disturbance. Asceticism.

The stately disturbance. Political quietism.

The matrimonial disturbance. Adopted by sons.

The need of a character portrait.

Reasons: a. a secret aversion against Democritus.
b. ignorance about the materialists.
(BAW 3 346).

4. On Democritus, Socrates and Plato: The Legend of Burning Texts

Nietzsche's longstanding hostility to the history of Plato and Platonism is well known. By his final years of sanity he notes that Plato is "pre-existently Christian," a claim that views Christianity as Platonism for the masses. The view that the body is a prison-house for the immortal soul has led many thinkers to desire to escape the body. Nietzsche has become famous for attempting to read metaphysical doctrines as symptomologies, to read the abhorrence of the body as an expression of physiological decline and weariness, a means to slander this world.[52] More importantly, Nietzsche seems to interpret (as Feuerbach and Marx had) the desire for an "other-world," whether it take the form of God, immortality, or the world of Platonic forms as a desire to escape the real conditions of existence, a psychological crutch to deny the tragic meaningless-ness of suffering on Earth. Plato's assessment of art as twice removed from the form (*eidos*) in his ontology, as well as his claim that "the Good" is "one" led Nietzsche to accuse Plato of confusing the least real with the most real.[53] Yet Nietzsche's hostility to the philosophy of Plato appears to have developed even earlier with regard to the alleged conflict between Plato and Democritus, as is evidenced by his writings from 1866 to 1867.

One might provisionally refer to the clash between Plato and Democritus as one between the natural and the supernatural (materialism vs. idealism, the physical vs. the metaphysical) aspects of their respective philosophies. While Democritus does argue that one should value the goods of the soul over the body, he nonetheless suggests that the soul, like the body, is fated to decompose.[54] Among other reasons, he appears as a philosophic hero to the young Nietzsche because he is not compelled to make a jump to supernaturalism.

Among other topics, Nietzsche's Diogenes Laertius investigations of the mid-1860s address rumors of philosophers trying to burn each other's works in the ancient world, a framework of investigation that identifies extra-textual conditions as crucial for the proliferation of any philosophic dialogue; i.e., Nietzsche sets out to inquire into the political and physical conditions necessary for the proliferation of philosophic works in the ancient world. In *The Birth of Tragedy* the presence of the doxographic Diogenes Laertius lingers: the story of Euripides being influenced by Socrates was used by Nietzsche to help develop his psychological critique of "aesthetic Socratism." This is the legend of Socrates agreeing to take Plato in as a student around 411 BCE, but only after Plato agrees to burn his tragic works (BT 14). Nietzsche uses this story to present a caricature of Socrates, characterized by a naive mania for dialectical justification, and hostility to the irrational elements (frenzy, excess, pain, annihilation, ecstasy) of tragedy. Nietzsche ultimately views tragedy as more fundamental than the ethical or epistemological worlds that gave birth to the Socratic ideal that views virtue, happiness, and reason as a natural trinity. Plato is supposed to have fallen as victim to the spell of Socrates, as Diogenes Laertius informs us that Plato had wrote dithyrambs and even had associated himself with the Heracliteans when he was young, before he would be "corrupted" by Socrates (DL III 5).

Nietzsche repeatedly demonstrates an affinity to Heraclitus of Ephesus in his philosophic orientation throughout his writings. Heraclitus, who supposedly held that one cannot step in the same river twice, is the fluxist who maintains that everything changes. In *Teleology Since Kant*, Nietzsche advances a Heraclitean view, suggesting that identities are inventions of the human understanding, existing only as surface representations. Although Plato rejects the position that all things are in a state of flux, Plato still agrees with Heraclitus about the nature of the physical world, insofar as it is subject to change. In Nietzsche's view, the world of change offers little hope for a metaphysical reward and punishment system in which the virtuous person is guaranteed happiness. Nietzsche's interpretation of the Heraclitean view is important for the understanding of tragic thinking, since it accentuates how the human understanding creates anthropomorphic identities, projecting them onto the river of flux which we call "world."

Ultimately Nietzsche presents the view that tragedy is deeper than any moral or teleological view of nature, so Socrates' peculiar moralizing tendency amounts to an optimistic escapism authored by a hyperactive discursive faculty. In this view, aesthetic Socratism is only made present by the human understanding in order to fathom a rational purpose for existence when no such purpose exists within nature. Tragedy reveals the conditional nature of the ethical dimension. As an art form tragedy is capable of depicting the nature of existence more truthfully than any moralizing discourse, since it shows the world as it is, without nostalgia for metaphysical reward and punishment systems. "Bad things" happen to "good people" in tragedy, since there is no moral world order in nature.

Nietzsche suggests that there are both ideological and epistemological optimisms invented in aesthetic Socratism. Socratism is a position that assumes that the world is a rational place, accessible to and correctable by discursive thought.[55] Nietzsche's caricature of Socrates in *The Birth of Tragedy* shares some of the features of Voltaire's Pangloss. Both Pangloss and Socrates wind up deceiving themselves about the significance and domain of discursive thought, each evidencing a naive faith in human reason.

Nietzsche employs the seemingly trivial story of Plato burning his own tragic works to develop a caricature of Socrates as a dialectical monster. The drastic measure of tossing texts into the fire reminds one that there are extra-textual conditions that decided the fate of whether a work survives or not. Nietzsche had been struck by Lange's powerful inquiry into the history of materialism that focuses precisely on this issue in Athens with Protagoras. Because Protagoras had denied that there was evidence for the existence of gods, his works were collected and burnt in public. (DL IX 51-2). The fact that inquiry into the nature of reality in ancient, medieval, modern, and contemporary times has often been policed and censored by unphilosophical ulterior motives remains an important intellectual issue, since this issue is frequently subverted, ignored, or denied by the ruling establishments.

Nietzsche's assessment of Plato's Socrates as a peculiar psychological type that hopes to formulate principles of a moral world order in *The Birth of Tragedy* may be a caricature, but such a portrait helps clarify the tension between aesthetic and moral views of the world. Mejer points out that the psychobiographical method that Nietzsche employs at times seems to suggest Diogenes Laertius's influence, since Diogenes Laertius also employed personal anecdotes to clarify, elucidate, and expound upon the characters of individual thinkers. Yet these are not the only stories of persons attempting to burn the works of philosophers on which Nietzsche focuses during his early years.

There is another legend of burning texts that preoccupied Nietzsche during his Democritus investigations, one that also involves Plato. Although receiving less attention, Nietzsche refers to it in multiple drafts: the rumor of Plato expressing the desire to burn all of the works of Democritus. In spite of Democritus being the most prolific author and most widely traveled philosopher of the tragic age of the Greeks, he is not mentioned by Plato even once, a fact that escaped the attention of neither Diogenes Laertius, nor Nietzsche.

It is Diogenes Laertius who calls attention to the possible reason why Plato never refers to him:

> Aristoxenus in his *Historical Notes* affirms that Plato wished to burn all the writings of Democritus that he could collect, but that Amyclas and Clinias the Pythagoreans prevented him, saying that there was no advantage in doing so, for already the books were widely circulated. And there is clear evidence for this in the fact that Plato, who mentions almost all the early philosophers, never once alludes to Democritus, not even where it would be necessary to controvert him, obviously because he knew that he would have to match himself against the prince of all philosophers. (DL IX 40)

Friedrich Lange's inquiry into the history of materialism points out that this may be more than a rumor, and character typologies had still not been given of the earliest Western philosophers:

> However incredible such fanaticism may appear to us, it is quite consonant with the character of Plato; and as Diogenes' authority for this statement is no less a person than Aristoxenos, it may be that we have here something more than a "story."[56]

Nietzsche's investigations of 1866-8 reveal the extent of the influence of Lange, not simply in regards to this particular rumor, but in Nietzsche's psychological analyses of philosophers. Roughly a year and a half after Nietzsche sent a letter recommending Lange's work to Gersdorff (August 1866), Nietzsche sends Gersdorff yet another letter commending Lange. In the latter letter of February 16, 1868 Nietzsche announces to Gersdorff that he intends to send his Democritus work to Lange as a measure of thanks when it comes to fruition. When Nietzsche was searching for a doctoral dissertation topic later that year in early May 1868, he writes to Rohde that both his Democritea and Homerica are "too good" to be used for the purpose of a dissertation topic. Although Nietzsche states that both of these themes are "too long-winded and too German" for dissertation purposes, one can still detect the lingering presence of Democritus in the later dissertation draft of *Teleology Since Kant*. Chapter three of this investigation will further explore this Democritean thread that connects Nietzsche to Lange's neo-Kantian philosophy.

In the early draft that follows, which is the final draft included in this first chapter of our investigation, Nietzsche announces a termination point of his early Democritus investigations. A doxographic collection has been left behind which is unsatisfying, for one cannot help but wonder what the totality of Democritus's original works were like, a collection that was unrivaled in his day. Nietzsche identifies a values clash at the basis of the destruction of Democritus's work: the philosophy of Democritus was believed to be unworthy from the perspective of a tradition in which Platonism and the drive to supernatural fanaticism triumphed.[57] Nietzsche experiences a sense of painful urgency in his desire to have something that is unattainable, a full recovery of the health and legendary brilliance of Democritus.

5. The Fate of Democritus (Short Translation)

Character typologies still have not been given for the ancient philosophers, i.e., Anaxagoras and Democritus and others are blurred together.

What has led us to undervalue Democritus? His determined opposition to teleology has. For Socrates, the lecture of Anaxagoras was the first epoch-making event in his life. He apprehended this point, found [Anaxagoras's] explanation of [teleology] to be bad, and could not help himself. Then came Socrates' second sailing [deuteros plous].[58]

It is to be noticed how Democritus distanced himself from the dissimulation. This is shown by the genuineness of his ethical writings: these always bid the corrective against malicious accusations. Among the ancients, the viewing of suicide is without reproach.[59]

Bad things have happened to the writings of Democritus: although they would be characterized as full of insightful judgments, as stylistic beauties, as model writings in a philosophic presentation, they would nonetheless be destroyed because in later centuries their justification would be felt as more and more strange, and especially by Christianity as it discarded the grounds for comprehending Democritus, as Aristotle had taken exception to his rejection of teleology. All but the hardest fate had already caught up to them a half century after the death of their composer: and this is truly the reason that the Christian scholars and monastic transcribers forced their hands from Democritus, to remove him as if he were possessed, a plan which Plato had kindled, to throw the collected writings of Democritus into the fire.

The drive for knowledge. [*aitiologia*] Journey.
Clarity: aversion against the wrong-headed.
Simplicity of method.
Purity of method.
Poetic swing (poetry in atomism).
The feeling of powerful progress.
Unconditional faith in his system.
Evils lie outside of his system.
Peace of mind as the result of the scientific investigation.
Pythagoras . . .

I can no longer pursue investigation about the lost writings of Democritus so as to feed my appetite with nourishment, since they are irretrievably lost: thus I am hardly half able to attain what I had sought after. We are guilty of a death sacrifice of Democritus, so that we can finally silence his angry spirit. And thus for whom would such health become other than for Democritus? Is his life not a martyrdom for science?

We are still very much guilty of the death sacrifice of Democritus, and only to some extent have we made good on the indebtedness to him by the past. In reality there was scarcely a more meaningful writer this diverse who had to suffer attacks whose origins sprang up from the most different motives. Theologians and metaphysicians had heaped their deeply rooted resentment (*Groll*) against materialism; the divine Plato held that his writings were so dangerous that he thought they should be annihilated in a private autodafé and only through reflection would he be dissuaded, that it would already have been too late, since the poison itself would already have been disseminated too far. Later, the obscurantists of antiquity would avenge themselves upon him, while smuggling their magical and alchemist writers into the establishment, and by doing so left fallow the father of all enlightenment and rational tendencies in the call for a great wise man. The setting in of Christianity finally achieved its follow through on the energetic plan of Plato: and certainly the anti-cosmological writings of the ancient Democritus, like the writings of Epicurus, appeared as the incarnation of heathenism. In our time, it finally remains that we also continue to deny the philosophic greatness of this man and he is recognized as a sophist. These attacks on Democritus are all set into motion from a basis which we can follow no further. On the contrary, we have an attack to meet them.
(BAW 3 345-7)

6. Recapitulation: Part 1

Friedrich Lange must be identified as a key influence on Nietzsche in 1866, especially in the exploration of the conflict between cultural mythologies and the development of the sciences. Although Nietzsche is sometimes viewed as a radical departure from the history of philosophy, such a view is frequently maintained by ignoring Nietzsche's philosophic mentors. Lange is difficult to overestimate in terms of his influence on Nietzsche and also proves to be important in Nietzsche's assimilation of the Kantian philosophy, a point for further investigation in chapter three. Nietzsche's methods of character analysis already begin to emerge within this context, along with a focus on teleology, elements of his philosophy that were in part stimulated by Lange.

Nietzsche's early reflections on the ancient Greek philosophers are also very likely shaped by the methods of Diogenes Laertius, as both find opportunities to connect personal details of a thinker's life to the individual thinker's philosophy. Although the whole of Diogenes Laertius is not satisfying as history, philosophy, or literature, his compilation captured Nietzsche's imagination and historical sense, furnishing Nietzsche a window through which to glimpse the ambiguous history of philosophy, before the setting in of the medieval world. Beyond the anecdotes that survive in Diogenes which are referred to both directly and indirectly in Nietzsche's analyses of the ancients, Diogenes offered Nietzsche the important lesson that the dissemination of philosophic texts is sometimes conditioned by political forces, as well as fear, jealousy, and spite. From Diogenes, we are offered a portrait of Democritus who shines above such pettiness, a model of health and wisdom for Nietzsche who was able to purify himself from superstition, greed, and vindictiveness.

Democritus thus appears as a crucial mentor for Nietzsche, a powerful thinker who philosophizes outside of the academy, not requiring any formalized school for validation. Yet this lack of a formal school may have contributed to the destruction of his works, for he would never be furnished with legions of transcribers in an era before moveable type, acid free-paper, or photocopying. As Nietzsche remarks in *Philosophy in the Tragic Age of the Greeks*, the survival of texts was a precarious situation in the ancient world. In general, philosophic texts needed to be recopied every few generations, usually on papyrus scrolls, a tedious and time-consuming task reserved for those who were literate. Nietzsche suggests that in later epochs the political and religious views of those who transcribed texts may have contributed directly to the destruction of the texts of Democritus, since the ancient atomist denies the premise held by virtually every world religion: the existence of the immortal soul.

Lange chronicles a history of philosophy that had rejected various kinds of idealism in favor of a mechanistic view of the world, an alternative to the long history of Platonism. Although it may be incorrect to refer to it as a history of atheism, the survey of materialistic thought could also be viewed in that light and it should surprise no one that such a work would be important for Nietzsche's assimilation of philosophers lying outside the ocean of Platonism. Just as

Copernicus had found forerunners in antiquity for his modern heliocentric view,[60] Nietzsche also finds an anticipation of his own views present in Democritus:

> Some men, not knowing about the dissolution of mortal nature, but acting on knowledge of the suffering in life, afflict the period of life with anxieties and fears, inventing false tales about the period after the end of life.[61]

Democritus's identification of human suffering as the psychological origin of false stories about the afterlife captured Nietzsche's attention, as it is a view that seems to anticipate Nietzsche's analysis of the origin of slave morality. Both thinkers identify the psychological cause of metaphysical punishment systems as coping mechanisms originating from pain, suggesting that the popular ghost stories of the underworld or hell can be detrimental to the well being of humanity. There is thus an affinity between Democritus and Nietzsche in that they both reject the ghost stories of antiquity as superstitious nonsense, suggesting that such stories originate from a peculiar animal that craves release from suffering. Mythology is interesting for psychological analysis, but it is rejected by both thinkers for the sake of eliminating the ideals of fear and suffering that negate the meaning of the good life for this world.

The exclusion of myth from the discourse of natural science proved to be crucial for the emergence of scientific methods, representing a conflict between Plato's teleological vision and the anti-teleological standpoint. By rejecting teleology, Democritus had left behind a theory of atomism that would be taken over by the Epicurean schools, one that proved later to be essential to the emergence of scientific methods in the modern era. Nietzsche was influenced by Lange's argument that Democritus's contribution to scientific methods was crucial for the development of modern sciences, but Nietzsche proceeds even further to claim that the ethical maxims formulated by Democritus are tragically neglected and most worthy of consideration. In Democritus Nietzsche had discerned a kindred spirit, a philosopher associated with laughter who was also a poet, free from academia, scholastic philosophies, spurious teleologies, and a model of self-sufficiency and wisdom. Such a figure provided an image of what philosophy could be, a hope for a philosophy from antiquity that could be in the service of life in the present.

Of the writings addressed in this investigation of Nietzsche's 1866 to 1868 philosophic reflections, the work on Democritus is the most difficult to assess, since Nietzsche tries out so many drafts on the history of the fate of the works of Democritus. Unlike Nietzsche's notes on Schopenhauer and Kant which immediately follow and overlap with the Democritus investigations, Nietzsche's Democritea does not offer a single unified thesis. Although Nietzsche's work on Democritus, Schopenhauer, and Kant from this early period focuses on teleology, Nietzsche's investigation of Democritus is different from his notes on Schopenhauer and Kant because Nietzsche valorizes the ethics of Democritus. Even though Nietzsche moves beyond any type of foundational atomism during this phase (as evidenced in *Teleology Since Kant*), he still recognizes the importance of the methods of atomism for rejection of teleological causes. The methods and the ethics of Democritus retain an important value for Nietzsche, an enduring vision that he could find in neither Kant nor Schopenhauer. Nonetheless, all three of these thinkers provide Nietzsche with important philosophic insights that shaped who he became as a philosopher.

The conflict between Platonism and Democritus may be expressed in terms of a conflict between "this-worldly" and "other-worldly" philosophies. Like Feuerbach and Marx, Nietzsche suspects that fixation on other worlds ultimately serves to deny and neglect the reality of this world, or even worse, slander this world. In this view, "other-worldly" philosophy takes the form of a metaphysical escapism. The Democritean ethics are heralded by Nietzsche precisely because they do not jump into the supernatural, favoring a sober scientific inquiry aimed at securing a strong, undisturbed disposition, free from bodily pain, anxiety, and disturbance. Nietzsche eventually terminated his Democritus project in a frustrated tone, suspecting that the Democritean writings had met a similar fate as the writings of Protagoras.

Although there is no way to corroborate the legend that Plato desired desperately to destroy Democritus's work for the sake of other-worldly metaphysics, Nietzsche at times appeared to accept that this rumor from Diogenes Laertius may have some authenticity. Plato's philosophical mission may be viewed as driven by underlying psychological causes, and the legend of burning texts would be indicative of a wider hostility to the Democritean view that the soul is fated to decompose, just as the body does. It may be the case that this rumor has no historical authenticity. No one knows. However, if nothing else, we may conclude that the conflict between Nietzsche and the history of Platonism runs deeply, even into the mid-1860s, where Nietzsche begins to make the transition from philology to a new kind of philosophy, and identify the conditions that determine whether texts will survive.

Notes

1. There is some love poetry attributed to Socrates from Diogenes Laertius's *Lives of Eminent Philosophers* (Cambridge: Harvard University Press, 1991) but the authenticity of such work is questionable. Precisely what to make of Diogenes Laertius's text is a theme to be addressed in this first chapter of our investigation.

2. This is addressed in Diogenes Laertius: DL (II 45) and DL (III 35-6). Such a view is not controversial, and indeed seems obvious given other aspects of the Platonic dialogues. R.G. Bury points out that Socrates is presented as being aware of the peace of Antalcidas in the *Menexenus* (387 BCE), over a decade after Socrates died. *Plato: Timaeus, Critias, Cleitophon, Menexenus, Epistles.* Translated by R.G. Bury (Cambridge: Loeb, 1999), 331.

3. Plato's *Timaeus* offers extended teleological explanations about the origin of life

and the universe that point directly to an intelligent designer. Whether such work is meant to be taken literally has been the subject of much dispute in the ancient, modern and contemporary worlds. Nietzsche gives a response to the *Timaeus*'s rambling account of intelligent design by presenting a sober response in *Teleology Since Kant*, a work addressed and translated in chapter three of our inquiry.

4. Nietzsche suggests that virtually all religions are deceptive about their own historical origins, as well as the nature of organic life and guilt, themes that are foreshadowed in his early philosophical investigations. In spite of Nietzsche's preoccupation with the Schopenhauerian theme of meaningless suffering, Nietzsche still distinguishes himself from Schopenhauer's gloomy pessimism by invoking a challenge to affirm life throughout his creative period.

5. Such a rumor does not seem inconsistent with Plato's views if one examines the *Laws*, Book 10, especially 908-9. Plato writes about imprisoning those who espoused that the world came to be by chance, and links this view to atheism. If atheists are released from prison and continue to espouse atheism, Plato suggests that they still should be put to death, even if they commit no crime other than advocating atheism. Nietzsche explores the conflict between the views of Plato and Democritus in detail, a conflict that will be addressed further in this phase of Nietzsche's genealogy.

6. For almost the entire twentieth century this period of Nietzsche's intellectual development was largely ignored by scholars. In part this was due to the omission of Nietzsche's late 1860s juvenilia from both the *Kritische Studienausgabe* and the *Kritische Gesamtausgabe*. The latter work has been amended to include these writings only at the very end of the twentieth century. James Porter's book *Nietzsche and the Philology of the Future* (Stanford, Calif.: Stanford University Press, 2000) is one of the only works to explore Nietzsche's encounter with Democritus in detail. Porter also examines some of Nietzsche's other 1860s philosophic work in *The Invention of Dionysus* (Stanford: Stanford University Press, 2000).

7. See Aristotle's *Physics* (2.2, 2.8) and *On the Parts of Animals* (1.1) for his criticism of Democritus's ruling out of teleological causes.

8. *Selected Letters of Friedrich Nietzsche.* Edited and translated by Christopher Middleton (Chicago: University of Chicago Press, 1969), 18. (BW 517).

9. Porter avoids the problem of trying to make pronouncements on this work by claiming that "Nietzsche has no single view of antiquity, and certainty no consistent view of it, at any point in his career. I want to suggest that he in fact has no views of antiquity" Porter, *Nietzsche and Philology*, 225. Although Porter's strategy of reading is methodologically useful for making sense (or not making sense) of the traditional hermeneutic project of coming to a unified view of what Nietzsche thinks, it seems to me that Nietzsche does hold some views about the meaning of antiquity, in spite of the contradictory theses he explores and advances. In fairness to Porter, he does offer a very good investigation of Nietzsche's focus on the recuperative limits of historical studies of antiquity, showing how the Democritus work overlaps with Nietzsche's Diogenes Laertius studies and general plans for a history of ancient philosophy.

10. Nietzsche considers this to be important in investigating the origin of language in his work on Diogenes Laertius. In DL, Epicurus expresses the strange view that "the names of things were not originally due to convention, but in several tribes under the impulse of special feelings and special presentations of sense primitive man uttered special cries. The air thus emitted was moulded by their individual feelings or sense-

differently according the difference of the regions which the tribes inhabited. Subsequently whole tribes adopted their own special names, in order that their communications might be less ambiguous to each other and more briefly expressed. And as for things not visible, so far as those who were conscious of them tried to introduce any such notion, they put in circulation certain names for them, either sounds which they were instinctively compelled to utter or which they selected by reason on analogy according to the most general cause there can be for expressing oneself in such a way." (DL X 75-6). This view is contrasted with Heraclitus, Democritus, and Aristotle, who derived language from convention, a view which Nietzsche took seriously in his analysis of the role language plays in representing the world and making claims about "what is."

11. André Ariew shows that the popular medieval view of teleology that commits oneself to nonmaterial forms is actually more Platonic than Aristotelian, especially when considering Plato's *Timaeus*. Aristotle is not nearly as teleological as the medieval thinkers make him out to be. Ronald Polansky points out that Aristotle does not write of inter-nature teleologies (i.e., birds are here for us to eat) and he almost never writes about a global teleology; however, he does hint at an overall arrangement at 12.10 of the *Metaphysics*.

12. Will Durant, *The Story of Philosophy* (New York: Washington Square Press, 1961), 225.

13. Susan Neiman makes this point in her provocative work *Evil in Modern Thought*: "Works like *Zadig* and the *Dictionary* see wisdom behind all those incredible chains of events. Works like *the Lisbon Earthquake* and *Candide* do not. (Princeton: Princeton University Press, 2002), 139. Neiman's book offers an interesting alternative history of modern thought that explores the loss of any connection of moral to natural evils in modernity. This is important for both Nietzsche and Voltaire, since such a connection is presupposed in the teleological views of the world that these two thinkers undermine.

14. Voltaire satirizes Pangloss's Leibnizian claim that "we live in the best of all possible worlds." Pangloss infers this from the presupposition that there is a supremely wise and omnipotent being who created the universe, Voltaire's comic evaluation of the problem of evil.

15. Michael Behe's *Darwin's Black Box* (New York: Free Press, 1996) is one recent example that promotes such views. Typically the literature that attempts to defend such arguments is not subjected to the rigors of peer review. Each generation yields someone who is willing to repeat this argument, invariably spawned by some theological bias. Richard Dawkins's use of cumulative selection in *The Blind Watchmaker* provides a reasonable response to the biological variation of the design argument. For a detailed interdisciplinary review of the politics and scientific status of this problem, see the *Chronicle of Higher Education*, December 21, 2001.

16. Urs Heftrich has investigated this connection to some extent in "*Nietzsches Auseindersetzung Mit Der 'Kritik Der Aesthetischen Urteilskraft.'*" *Nietzsche Studien*. Bd. 20, 1991.

17. (TSK [BAW 3 375]).

18. In his excellent book *Imagination and Interpretation in Kant* (Chicago: University of Chicago Press, 1990), Rudolf Makkreel explores how the Kantian position evolves by the time Kant writes the *Critique of Judgment*, as Kant discloses powers of the imagination and the judging subject in that work which are not present in the *Critique*

of Pure Reason.

19. Nietzsche sketches out a sense of this view more eloquently in his later essay *Truth and Lie in a Non-moral Sense*, a view that suggests humans overestimate their own perspective, falsely conceiving themselves as having an important significance within the cosmos:

> In some remote corner of the universe that is poured out in countless flickering solar systems, there once was a star on which clever animals invented knowledge. That was the most arrogant and the most untruthful moment in "world history"—yet indeed only a moment. After nature had taken a few breaths, the star froze over and the clever animals had to die. Someone could invent such a fable and still not have illustrated adequately how pitiful, how shadowy and fleeting, how purposeless and arbitrary the human intellect appears within nature. There were eternities when it did not exist; and someday when it is no longer there, not much will have changed. For that intellect has no further mission leading beyond human life. It is utterly human, and only its owner and producer takes it with such pathos as if the whole world hinged upon it. But if we could communicate with the gnat, we would learn that it too swims through the air with this same pathos and feels within itself the flying center of this world. (TL 1)

20. Nietzsche claims at various junctures that "Every identity is imaginary" and "Being is an empty fiction." This second claim certainly can be viewed as a response to Plato's question from the *Timaeus*: "What is that which is existent always and has no becoming?" (27d). In Plato's view, the existent (being) has nothing to do with the world of sense perception, and this is reflected in his dictum that "as Being is to Becoming, so is Truth to Belief" (29c). Plato, *Timaeus, Critias, Cleitophon, Mexexenus, Epistles.* Translated by R.G. Bury (Cambridge, Mass.: Loeb, 1999). Plato advances a similar view toward the end of *Cratylus*. Nietzsche was attentive to and may have also been influenced by Diogenes Laertius's chapter on Plato, which provides some exposition of this problem. Diogenes Laertius writes: "Plato asserts that the object of sense is that which never abides in quality or quantity, but is ever in flux and change. The assumption is that the things from which you take away number are no longer equal nor determinate, nor have they quantity or quality. These are the things to which becoming always, and being, never belongs. But the object of thought is something constant from which nothing is subtracted, to which nothing is added. This is the nature of eternal things, the attribute of which is to be ever alike and the same" (DL III 9-10). This part of Diogenes Laertius's history of Plato is controversial, since it alleges that Plato took over the comic poet Epicharmus's view of sense perception. The authenticity of the fragments that suggest Plato stole some of his teachings from Epicharmus and others has been contested by Nietzsche's friend Rohde, Williamowitz, and others. See Long's comments in volume one of Diogenes Laertius, *Lives of the Eminent Philosophers.* Translated by R.D. Hicks, (Cambridge, Mass.: Harvard University Press, 1991), 284-5.

21. In its assumptions of an intelligent cosmic order, the *Timaeus* provides the most full blown version of a spurious teleology in the Platonic corpus, a view that goes far beyond Aristotle's observation of the order in nature.

22. Porter, *Nietzsche and Philology*, 124.

23. As Porter indicates, this view was already entertained in the ancient world, where the problem of what or who Homer was had been reflected upon by Cicero, Plutarch, Aelian, Josephus, and others.

24. Nietzsche's philosophy may be interpreted as a revolt against Platonism on at least four different counts, all of which are connected to the evaluation of teleology. Firstly Nietzsche rejects the desire to escape the body that is implicit in Plato's view of the body as prison-house to the soul. Secondly, Nietzsche rejects Plato's ontology that claims art is the least real. Thirdly, Nietzsche rejects the idea that the good is "one." A fourth contrast could be explored in terms of the origin and nature of both essences and ideas. Nietzsche's rejection of Plato could also be framed on the very basic issue to be explored here, a rejection of discursive anthropomorphic natural teleologies that suggest that the cosmos is ordered by an external designer.

25. However, as Greg Whitlock points out, Nietzsche's lectures from the early 1870s (the time of *PTA*) do emphasize the importance of Democritus, an emphasis that is in fact already present in the mid- to late 1860s: "Of all the more ancient systems, the Democritean is of the greatest consequence. The most rigorous necessity is presupposed in all things: there are no sudden or strange violations of nature's course. Now for the first time the collective, anthropomorphic, mythic view of the world has been overcome." Whitlock, *The Pre-Platonic Philosophers/Friedrich Nietzsche* (Chicago: University of Illinois Press, 2001), 125.

26. (BAW 3 349). In the final analysis Anaxagoras's claims about *Nous* directing the world are also free from anthropomorphic teleologies. In Nietzsche's later lectures on the pre-Platonics, he writes "Anaxagoras was far removed from a direct purposive end for all individual things, and this is the point where Plato (in the *Phaedo*) and Aristotle launch criticisms of him. . . . Anaxagoras wanted to explain the world with the fewest possible nonphysical theories. For him, circular motion suffices; had he immediately imagined an intellect with continual purposive ends, it would have been a mythological being, a god—precisely what he dismisses." Whitlock, *Pre-Platonic Philosophers*, 98-9.

27. For more details on Nietzsche's philological investigations of Diogenes Laertius, see Porter and Jørgen Mejer's work, *Diogenes Laertius and His Hellenistic Background* (Wiesbaden: Steiner, 1978). Mejer points out that in the late nineteenth century and early twentieth century it was proposed repeatedly that, "Diogenes copied primarily from one source with a few additions of his own. These theories were all refuted as soon as they were published (Nietzsche by Diels, Maas by Wilamowitz, Usener by Gercke)" (4-13).

28. See Porter's *Nietzsche and Philology* for explication of Nietzsche's philological projects during this phase, especially chapter two. Ritschl was one of Nietzsche's greatest teachers and remained an important encouraging force for Nietzsche as a professional philologist, supporting and advancing Nietzsche's career until the publication of the *Birth of Tragedy*. When Ritschl accepted a position at Leipzig in 1865 Nietzsche transferred from Bonn to Leipzig for the specific purpose of continuing his studies under Ritschl.

29. The rumor of Socrates directing the composition of Euripides' plays (DL II 18), as well as Nietzsche's description of the individuating principle as a ship on the ocean (DL IX 68), both come from Diogenes Laertius.

30. Jørgen Mejer also stresses this point. *Diogenes Laertius*, 61.

31. While much of his compilation is ignored, some of the defining work attributed to Epicurus that is routinely and widely accepted as authentic comes from Diogenes

Laertius's compilation, such as Epicurus's *Letter to Menoeceus*, etc.

32. The problem with the legend as it relates to Diogenes Laertius is summed up by Herbert S. Long in the preface of the Loeb edition: "Diogenes has acquired an importance out of all proportion to his merits because the loss of many primary sources and of the earlier secondary compilations has accidentally left him the chief continuous source for the history of Greek philosophy." Long, Preface to *Lives of the Eminent Philosophers* by Diogenes Laertius. Volume I (Cambridge, Mass: Harvard University Press), xix.

33. Regarding Diogenes Laertius, Long points out that "Richard Hope counted 1186 explicit references to 365 books by about 250 authors, as well as more than 350 anonymous references." Long, *Diogenes*, xix. Mejer explores the difficulty of this task and presents a persuasive case that the claim that Diogenes Laertius may have been dishonest and deliberately deceptive is unfounded and implausible.

34. This aversion is expressed well by Gary Alan Scott, who has referred to Diogenes Laertius as "the *National Enquirer* of the ancient world." Clearly not *everything* can be believed that is found in the compilation of Diogenes Laertius. However, it is intellectually indefensible to dismiss all of Diogenes as fictitious tabloid. Even if we employ all the best methods, there still are junctures that lack sufficient evidence to conclusively determine whether aspects of Diogenes Laertius are historically authentic. I refer to these aspects of Diogenes as "the legend." Some of "the legend" is employed by Nietzsche for some of his own psychological analyses of the characters of the early Greek philosophers. Nietzsche's lectures on the pre-Platonic philosophers, as well as BT and PTA, draw heavily from Diogenes Laertius.

35. See (DL VI 40-42, etc.) for the peculiar tension alleged to exist between "the Dog" and Plato.

36. Although he considers philosophic positions various philosophers held, Diogenes is more concerned with their life activities. What can be said about the lives of the philosophers takes precedence in his histories.

37. In chapter two of our inquiry, we have the benefit of examining Nietzsche's longest sustained philosophical analysis of the specific details of Schopenhauer's philosophy. Elizabeth Nietzsche, Claudia Crawford, and John Sallis correctly point out that the *Fragment of a Critique of the Schopenhauerian Philosophy* demonstrates that Nietzsche did not uncritically and naively adopt Schopenhauer's theory of will in a so-called early phase of Nietzsche's philosophy. Over the course of this investigation, it will be demonstrated that the secondary scholarship that refers to these divisions of Nietzsche's philosophy into various epistemic phases cannot be maintained when one examines Nietzsche's longest sustained philosophical analyses of Schopenhauer and Kant in 1867-8.

38. Neiman accentuates the irony of how Kant, Hegel, and Wittgenstein all are determined to avoid the "merely" psychological. She writes, "the hostility toward psychologism in a field concerned with self-knowledge is deep and striking. For all the great differences between them, one thing uniting Rousseau and Nietzsche is disdain for other philosophers' distinction between philosophy and psychology." Neiman, *Problem of Evil*. 212. Her general view that the relationship between philosophy and psychology is now more than ever in need of clarification seems well founded.

39. (BAW 4 213).

40. Elizabeth Nietzsche and Pletsch explore this issue by viewing Schopenhauer as a

father figure for whom Nietzsche yearns. Nietzsche does exhibit an obsessive admiration of Schopenhauer's character in the early years, even though he recognizes faults in Schopenhauer's system.

41. "'As to the gods, I have no means of knowing either that they exist or that they do not exist. For many are the obstacles that impede knowledge, both the obscurity of the question and the shortness of human life.' For this introduction the Athenians expelled him; and they burnt his works in the market-place, after sending round a herald to collect them from all who had copies in their possession" (DL IX 51-2).

42. Travel was obviously much more dangerous and difficult during the fifth century BCE. Some of Diogenes' report of the "history" of Democritus's travels and training under Persian Magi could be apocryphal:

> According to Demetrius in his book on *Men of the Same Name* and Antitsthenes in his *Successions of Philosophers*, he [Democritus] traveled into Egypt to learn geometry from the priests, and he also went into Persia to visit the Chaldaeans as well as to the Red Sea. Some say that he associated with the Gymnosophists [the naked philosophers] in India and went to Aethiopia. (DL IX 35)

43. Nietzsche notes that "The origins of Pyrrhonism and Epicureanism lie in Democritus. The former out of a principle pertaining to knowledge, the latter from the ethical point of view" (BAW 3 332). Democritus's claim "In reality we know nothing, for truth lies in an abyss" (DL IX 72) captured the attention of both Pyrrho and Nietzsche.

44. Gregory Vlastos. "Ethics and Physics of Democritus." In *Studies in Greek Philosophy I* (Princeton, N.J.: Princeton University Press, 1995), 328-50. Porter has done some work connecting Vlastos and Nietzsche, suggesting that Nietzsche anticipates Vlastos's groundbreaking work linking the Democritean ethics and physics nearly a century before Vlastos.

45. *Metriopathia* indicates a sense of due measure in the emotions or passions.

46. Nietzsche himself notes in his reading of Diogenes Laertius that Archelaus as well as Democritus had also distinguished themselves as ethical thinkers before and during the time of Socrates. Thus the view that Socrates is a turning point in the history of philosophy in his exclusive focus on the ethical dimension may be overstated. Archelaus is also alleged to have turned his attention completely to ethics. Unlike Archelaus or Democritus, Plato had a school devoted to promoting and furthering Socrates' philosophic reflections.

47. Marx writes: "The skeptical, uncertain and internally self-contradictory view held by Democritus is only further developed in the way in which the relationship between the atom and the world which is apparent to the senses is determined." Karl Marx and Friedrich Engels. *Collected Works*, vol. I. Translated by Dirk J. and Sally R. Struik (New York: International Publishers, 1975), 39. Even if there is a type of skepticism built into the Democritean worldview, there is also a mandate for action in the surviving fragments, evidenced in Democritus's numerous claims that action is more important and significant than speech.

48. In his later lectures on the pre-Platonic philosophers, Nietzsche points out that the belief that Democritus used to laugh at all things originates later. See Whitlock, *Pre-Platonics*, 122.

49. (BAW 5 305).

50. Nietzsche provides similar comments at (BAW 4 44) regarding the neglect of Democritus.

51. Democritus writes: "(I would) rather discover one cause than gain the kingdom of Persia." Fragment #18, *Ancilla to the Pre-Socratic Philosophers*. Translated by Kathleen Freeman. (Boston: Harvard University Press, 1948) [hereinafter Democritus frag.—Freeman].

52. See (GS Pref II) and (TI III 6) for the theme of interpreting otherworldly philosophy as life-negating escapism. The rejection of the strange desire to escape the body appears as a theme at multiple junctures in Nietzsche's thinking, a common factor that underlies his evaluation of the ascetic ideal, and appears in Zarathustra's oracular mandate to "remain faithful to the earth."

53. Later Nietzsche suggests that there are multiple histories of the concept "Good," conditioned by whether the valuators are powerful or powerless, unlike Plato's mysterious concept of "the Good," which appears as a single reality principle, one that is not invented by human beings.

54. Democritus does not exhibit any desire to blame the body as a source of corruption. He writes, "If the body brought a suit against the soul, for all the pains it had endured throughout life, and the ill-treatment, and I were to be the judge of the suit, I would gladly condemn the soul, in that it had partly ruined the body by its neglect and dissolved it with bouts of drunkenness, and partly destroyed it and tore it in pieces with its passion for pleasure—as if, when a tool or a vessel were in bad condition, I blamed the man who was using it carelessly." Democritus frag. 159 Freeman. This sounds like the maxim "a good carpenter doesn't blame his tools." However in this case, a good soul doesn't blame the body. Democritus held the soul is fated to decompose, but he still offers ethical advice in the surviving fragments that stress the importance of action and virtue, while rejecting pettiness, empty talk, meanness, vindictiveness, and greed. Nietzsche thinks that Democritus had been ignored due to the tradition of Platonism, one which is at odds with the Democritean worldview. Moreover, Democritus's claim that "One should tell the truth, not speak at length" further indicates a type of integrity that may be foreign to a culture of Socratism and modern scholastic pedagogy. Democritus frag. 225 Freeman.

55. Even if Diogenes Laertius's historical rumors of Socrates helping Euripides compose tragic works are apocryphal, the point is that this is a tendency found in a type of moral theorizing. Nietzsche writes that "Socratism is older than Socrates" at (KSA 1 545).

56. Lange, Friedrich. *History of Materialism*. Translated by Ernest Chester Thomas (Boston: Houghton, Osgood & Co, 1880), 18.

57. The Democritean writings in question were lost after Thrasyllus, probably in the first century under the reign of Tiberius. However, Nietzsche suggests that there is a larger need to recognize that the historical survival of texts has been conditioned by the value judgments of those who transmit and copy texts, for such judgments function to advance or repress their survival.

58. This is Socrates' second voyage, a search for a teleological explanation of the world. As indicated in the *Phaedo* (97c-99d), Socrates had sought out Anaxagoras's *Nous* as an explanatory principle, only to become disappointed with Anaxagoras's minimalist account of it as a director of the cosmos. In Nietzsche's later lectures on the

pre-Platonic philosophers, he writes, "Well now, the actual world reveals itself to us not as a chaos but instead as order and beauty, determinant lawfulness, and so on. Chance, Anaxagoras says, cannot explain such things. What is it then, that so orders and arranges lawful regularity? . . . This is the important idea of Anaxagoras, that rotation suffices to explain all order and regularity in the universe. . . . We should not, then, confuse him, without further qualifications, with the teleologists." Whitlock, *Pre-Platonic Philosophers*, 97-8.

59. Democritus is supposed to have taken his own life in old age when he was somewhere between 90 and 110 years old. By the time of his suicide he was supposed to be blind and experiencing digestive failure. Like Plato and Socrates, Democritus also claims there are certain conditions under which life is not living: "Life is not worth living for the man who has not even one good friend" Democritus frag. 99 Freeman. Nietzsche notes to himself how the ancients view suicide without reproach when taken at the right times and conditions, very much in contrast to the view taken by the teleological Platonism of Aquinas, where suicide is viewed as contrary to the purposive design of God. Indeed most of the contemporary Western countries forbid suicide and make it illegal. This is due partly to religious beliefs shaping popular morality in the West. Only recently have there been any organized groups in the West that try to legalize suicide, such as the Hemlock Society (referring to Socrates) in the United States and Dignitas in Switzerland. At this time suicide is legal under clearly specified conditions in the Netherlands, as well as in the state of Oregon. This issue raises the important question of whether the ancient philosophers are capable of providing us any insight for important decisions in the present day, a key theme for Nietzsche, as history can be used or abused as a mode of instruction for the present.

60. Copernicus writes, "I undertook the task of rereading the books of all the philosophers I could get access to, to see whether any one ever was of the opinion that the motions of the celestial bodies were other than those postulated by the men who taught mathematics in the schools. And I found first, indeed, in Cicero that Nicetas perceived that the Earth moved and afterward in Plutarch I found that some others were of this opinion." N. Copernicus. "The Revolutions of Heavenly Bodies." In *The Philosophy of the 16th and 17th Centuries*. Edited by Richard H. Popkin (New York: Free Press, 1966), 49. Nietzsche at times compares himself to Copernicus, even going so far as to conceive himself as a Pole. They both have the effect of displacing the human species from a privileged place in the center of the universe.

61. Democritus frag. 297 Freeman.

Chapter Two

Nietzsche on Schopenhauer in 1867

1. The Phenomenal Bridge

This second chapter of our inquiry addresses Schopenhauer as Nietzsche's philosophical mentor. This investigation of Nietzsche is organized in terms of a chronological progression from the late 1860s, and this second chapter focuses primarily on Nietzsche's 1867 *Fragment of a Critique of the Schopenhauerian Philosophy* (FKSP). Although various scholars have asserted that Nietzsche uncritically adopted Schopenhauer's theory of the will in the so-called early phase of his philosophical development, the fragment from 1867 indicates that Nietzsche specifically rejected Schopenhauer's claim to know the thing-in-itself on fairly rigorous philosophic grounds in the late 1860s.[1] Moreover, chapters two and three of this investigation indicate that there are serious problems with some of the scholarly literature that has introduced divisions that isolate a perspectival stage of Nietzsche's thought as well.

Examination of Nietzsche's early analyses indicates the clear presence of an aesthetic perspectivism during the 1860s. Nietzsche consistently interprets the claims to truth in philosophy as a function of an all-too-human interpretive viewpoint. He consistently rejects the existence of any supernatural truths and is unconditionally suspicious about claims to knowledge that purport to offer a view that completely suspends all human perspective. Nietzsche synthesizes a radical philological historicism with a view that takes seriously the modern critiques of representation present in Kant and Schopenhauer. Such critiques had stressed the limited function of the human intellect, a theme which Nietzsche explores during this early phase, as well as throughout his mature works. The significance of human consciousness within the species and the cosmos is a theme that recurs throughout Nietzsche's corpus. Such issues are important in and of themselves, but they are also significant for understanding the nature of humanity's own genealogy. These issues also have direct implications for the classic philosophic quest for knowledge of self. The meaning of human life for the species and for the individual is not a nostalgic relic of pre-twenty-first century thinking, but persists as a genuine philosophic concern.

Nietzsche's philosophical reflections from 1867 to 1868 examine the degree to which human beings can think the world *as it is*. By probing the limits of the human understanding, Nietzsche stresses the anthropomorphic nature of human cognition. The FKSP and Nietzsche's critique of teleology in the following year (TSK 1868) furnish us with the longest sustained epistemological analysis of Schopenhauer and Kant from the entire Nietzschean corpus, so these early fragmentary works are clearly important for consideration to assess Nietzsche's development and relationship to these thinkers.[2] Moreover, they furnish us with fairly unambiguous critiques, free from the ambiguities that are so often associated with Nietzsche's literary excesses. During this phase, Nietzsche already articulates serious reservations with the drive for system-building in philosophy (FKSP 2).

In the aftermath of Kant in the nineteenth century, there was a widespread view among many German idealists that it was essential to have a system for serious philosophical analysis. In this respect, there are some superficial resemblances of Kierkegaard to Nietzsche in terms of their rejection of the necessity of philosophizing within a system.[3] Nietzsche's early interpretation of Schopenhauer in FKSP appears less dogmatic than his later rejection of systematicity, as he provides an account in 1867 that specifically analyzes the problems of systematicity in the larger context of Schopenhauer's critique of representation.[4] In FKSP Nietzsche provides specific analysis of Schopenhauer's arguments to show that the Schopenhauerian theory of the will is contradictory.

In 1867, Nietzsche already begins to explore the limits of discursive representation and the implications such limits have for assessing and evaluating the meaning of truth in philosophy. The nature and scope of Nietzsche's analysis suggests that there is already a mature perspectivism present in this early work. FKSP shares some of the same areas of inquiry with TSK, as both are written during a phase of biological reflection in which Nietzsche considers the implications of Darwin's theory of the origin and preservation of life that appeared less than a decade earlier. Nietzsche had noted in his Democritean reflections that the world and life itself need not have any intelligent designer, a theme that he develops at length in his wholesale rejection of teleology in chapter three of our inquiry. Such reflections clearly evidence the deep influence of Friedrich Lange on Nietzsche's early thinking about biology, as Lange specifically addresses Darwinism and the problem of teleological consideration of organic life in the second volume of the *History of Materialism*.

Schopenhauer advances his own biological reflections, focusing on a different kind of epistemological limit that amplifies and deviates from Kant's critique of the limits of human knowledge. Although accepting much of Kant's project, Schopenhauer offers a different exploration of the individual within the context of the species, and presents a very strange alternative to the Kantian thing-in-itself. Kant's investigation of the organism in the *Critique of Judgment* provides Schopenhauer with a launching point for his own philosophy, since there is in Kant's view something epistemologically inaccessible about the organism. In a sense, Schopenhauer interprets the life force present within the organism as a continuation of Kant's explication of the noumena from the *Critique of Pure Reason*, as both the noumena and the natural organism indicate a region that goes beyond the epistemological limitations of the knowing subject.

Kant's Copernican revolution of the *Critique of Pure Reason* introduced a new concept of nature, a nature that was a function of the forms of sensibility and the categories of the mind through which the knowing subject has experience. Although Schopenhauer accepts much of Kant's analysis regarding the active role of the subject which structures phenomenal experience in space and time, Schopenhauer departs from the Kantian view and its attempt to understand the living organism. For Schopenhauer, the living organism is the condition for the possibility of discursive reflection, an acknowledgment that there is an important sense of "nature" for which Kant did not account in the *Critique of Pure Reason*. Kant's radical new sense of nature from the *Critique of Pure Reason* (nature as a function of an active structuring by the transcendental subject) leaves out the other sense of nature that is more common and often associated with a primitive world that is free from constructions of the human mind.[5] However, this traditional sense of nature erupts through Kant's concept of the genius in the *Critique of Judgment*, an active rather than passive nature, one that does not appear to be merely a function of a transcendental subject's form-giving structure. This traditional sense of original nature is crucial and primary for Schopenhauer, who names the underlying reality of a dark pre-discursive force of the will as the "thing-in-itself." This is why Schopenhauer views the human intellect as a contingent faculty and region of representation built on (and made possible by) the underlying existence of a "natural" will:

> It is not the intellect that has produced nature, but nature that has produced the intellect.[6]

During the 1860s, the status of the Kantian and Schopenhauerian thing-in-itself animates Nietzsche's surviving philosophic notes. There does indeed seem to be an eruption of nature through the subject that appears in Kant's *Critique of Judgment* and the Schopenhauerian will, one that may suggest a conditional existence of the transcendental subject. Even though Kant never fully explored the philosophical significance of this dimension which suggests that the self is a contingent structure, Kant's description of the shuddering subject in the analytic of the sublime opened a new direction for philosophic thinking in the modern and contemporary world. The awesome power of nature fuels Schopenhauer's thoughts on the dark unconscious will and finds a secondary expression in Nietzsche's own description of nature in the Dionysian. There does seem to be an eruption of "original nature" in the Kantian sublime, Schopenhauer's will, and Nietzsche's Dionysian which suggest that the knowing subject is vicarious, conditional, and contingent in nature—even if Kant would never draw or accept this conclusion. The knowing subject as individuating principle could be compared to a boat tossed by a violent storm on the ocean, an image that

Nietzsche retrieves from Schopenhauer (BT 1), originating in Diogenes Laertius's chapter on Pyrrho (DL IX 68).

Kant, Schopenhauer, and Nietzsche describe types of experiences that challenge the limits of the knowing subject. These experiences are philosophically important because they may help jar the knowing subject to become aware of its own finitude, a contingent finitude that is borrowed from a larger unbounded power.[7] The Kantian sublime, Schopenhauerian will, and Nietzsche's Dionysian all reveal a feeling of smallness within the individual knowing subject when it compares itself to the awesome might of nature. Kant does not draw the types of radical inferences that Schopenhauer and Nietzsche do, but his third critique tests the limits of the human understanding in ways that were not explored in the *Critique of Pure Reason*, both in terms of the aesthetic presentation of the unbounded, as well as in the context of teleological judgments, a theme to be explored further in chapter three of our inquiry.

Even when Nietzsche reiterates a rejection of Schopenhauer's theory of the will in the later writings (BGE 19), one cannot explain away the striking philosophic debt to Schopenhauer, especially in terms of the role and status of the human intellect. Schopenhauer suggests that the human intellect is something that appears fairly recently within the eons of time, authored by an image-creating power. The chance origin of the human intellect that is assumed by Schopenhauer's analysis of the phenomenal world clearly influenced Nietzsche's thoughts on the Apollonian, and Schopenhauer's attempt to present his thoughts on the will influenced Nietzsche's concept of the Dionysian in the *Birth of Tragedy*.[8]

Even though Schopenhauer claims to know the thing-in-itself, he still retains an important allegiance to the spirit of Kant's critical enterprise, one that generally leaves humans immersed in the world of appearances. However, that is only half of Schopenhauer's story, since original nature at times breaks through the representational faculties, revealing a different, "other" type of experience which seems unmediated and largely free of discursivity. Although not used consistently within Schopenhauer's system, this dimension identifies the animal force within human beings, an important underlying principle of life that had not been adequately explicated within the history of philosophy's self-reflexive intellectual representation.

Schopenhauer clearly exerted a profound impact on Nietzsche by his analysis of the dark unconscious will and its connection to the significance of music. This is especially evident in the *Birth of Tragedy*, where Nietzsche refers to Dionysian music as a kind of way to access the thing-in-itself.

> For, as we have said, music is distinguished from all other arts by the fact that it is not a copy of the phenomenon or, more accurately, of the adequate objectivity of the will, but an immediate copy of the will itself, and therefore complements everything physical in the world and every phenomenon by representing what is metaphysical, the thing-in-itself. (BT 16)

Of all of Nietzsche's published works, the *Birth of Tragedy* is the only work that employs and seems to endorse the meaning of the thing-in-itself as a glimpse into the unconditioned. In all other places Nietzsche interprets the thing-in-itself as a fiction, suggesting that the thing-in-itself amounts to a view without perspective, a rhetorical sham, or an outright contradiction. Later Nietzsche would look back in embarrassment and regret of this phase in his later preface to the *Birth of Tragedy*, since he had fumbled around with Kantian and Schopenhauerian jargon, straining their meanings and endorsing a youthful abandonment of sobriety in his exploration of the significance of music.[9]

Ultimately Nietzsche's *Birth of Tragedy* was searching for a means to articulate his deep admiration for the sublimity of music, dubbing the special experience of Dionysian music as a way to the noumena. It would be a mistake to discount Wagner's influence on Nietzsche during the late 1860s and early 1870s, since Nietzsche had fallen into a subservient role to Wagner's egomania shortly before the *Birth of Tragedy*'s creation. Nietzsche first met Wagner in November 1868, several years prior to the composition of the *Birth of Tragedy*. In their first meeting, Wagner and Nietzsche found opportunity to discuss Schopenhauer's theory of music, a common personal interest that helped in creating the bond that blossomed into their friendship.[10] Wagner seems to have been psychologically committed at times to the Schopenhauerian idea that the composer has a special connection to the underlying reality of the immediate language of the will.[11] Such a theory inflated the significance of the composer, suggesting a need to reevaluate the cultural meaning of the composer, as well as a need to reevaluate general suppositions about the relationship of music to expression, representation, and truth.

Just as one might describe a symphony as "divine" without really implying that it has anything to do with gods, Nietzsche describes a different mystical place in the *Birth of Tragedy*, afforded by a spirit of life-affirmation rooted in Dionysian revelry. Nietzsche's description of the experience of the loss of self that overtakes and negates the barriers of the individual purports to steal a glimpse into an underlying oneness with the world, an insight afforded by annihilating the separation of the transcendental subject from nature. Without explaining away Nietzsche's intentional and deliberate use of the language that describes such an occurrence in terms of Schopenhauer's theory of the will, one could still suggest that Nietzsche was simply trying to point to a dimension that cannot be conveyed adequately through discursivity and linguistic representation. In this sense, the experience of Dionysian dissonance crashes through the epistemological limits of discursive representation, yielding a type of experience of primordial oneness (*Ur-eine*) that makes the individual feel small in the context of nature. Dionysian music and sex are able to trigger the feeling that the individual is subservient to (and less significant than) a larger lifeforce that animates the species. Such a move is very close to Schopenhauer's theory of the will, a move that Nietzsche seems to reject on well argued theoretical grounds in FKSP. However, Nietzsche's primordial one is not Schopenhauer's will. Moreover, it is neither primordial nor one.

In the *Birth of Tragedy*, Nietzsche writes of the primordial oneness (*Ur-eine*) and the thing-in-itself in a very different sense and mood than his even earlier comments from 1870-1:

> There is no way to the primordial one for humans. It is completely appearance. (KSA 7 170)

> The primordial one reveals something as process, temporal, spatial, and causal. (KSA 7 186)

If the primordial one is "temporal, spatial, and causal," it still remains in the network of representation, an appearance in the phenomenal world. Later, in the *Birth of Tragedy*, the primordial one mysteriously appears again in Nietzsche's writing, just as the thing-in-itself does. Ultimately Nietzsche conveys a sense that it is possible to experience the *Ur-eine* and the thing-in-itself in the form of Dionysian excess in the *Birth of Tragedy*, a glimpse into a deeper reality that underlies individuation. As Clarke indicates in *Nietzsche on Truth and Philosophy*, the peculiarity of the *Birth of Tragedy* is due to Nietzsche's desire to give expression to such experiences by describing them in terms of the noumena. However, it is not the case that such a view indicates that Nietzsche held this as a general philosophic position prior to 1872. Rather, Nietzsche specifically rejects the Schopenhauerian leap to get to the noumena several years earlier, in the FKSP of 1867.

Even before his discovery of Schopenhauer's theory of music in late 1865, Nietzsche had presented numerous artistic projects involving music to his artistic-literary club Germania.[12] In September 1862, he had proposed an essay contest "On the Essence (Wesen) of Music," a theme that anticipates elements of his investigations of the *Birth of Tragedy*. Nietzsche was fascinated with the significance of music's uncanny and superpowerful effect on the listener, as evidenced in his notes (BAW 2 89, etc.) at least as early as 1862. He suggests during this time that music "symbolically represents the movement of the cosmos," an attempt to explicate the strange power that Nietzsche describes as something godly: a demonic force! Although he does not explicitly offer the *Birth of Tragedy*'s view that Dionysian music is a means for one to recognize the conditional nature of individuation through the experience of the *Ausser-sich-sein*, the practical and theoretical significance of music was incredibly important for Nietzsche throughout his life.[13]

Nietzsche felt compelled to investigate music in the context of nineteenth century critiques of representation and used the somewhat clumsy philosophical categories inherited from the tradition of Schopenhauer and Kant. Ultimately the *Birth of Tragedy* is about music and neither Schopenhauer nor Kant's terminus on the thing-in-itself adequately explicates the special types of rapture and life-affirmation that Nietzsche connects within Dionysian music. It is music that transgresses the discursive epistemological limit, revealing a force that Schopenhauer and Nietzsche identify as a mysterious power without parallel.[14]

Nietzsche attempts to use the inherited epistemological categories for the purposes of aesthetic critique, but the distinction between noumena and phenomena is strained and broken by Nietzsche's attempt to explicate the distinction between music and the nonmusical.

Schopenhauer's theory of music had dealt the composer a metaphysical role that had formerly been dominated by religion: a purported connection to the absolute. In this respect, both Wagner and Nietzsche viewed music differently from all other arts, an expression of a tonal art that was more fundamental than the words that constituted traditional philosophy. No amount of poetic eloquence can capture the significance of music, and Nietzsche stammers to express the inexpressible by crossing from the ordinary phenomenal world into the region he names as the Dionysian. Music, like the sex drive, appears as a dynamic power to Nietzsche that at times seems to occupy a completely separate world than the domain of speech and philosophy. The larger question of whether Nietzsche exchanges one fictitious claim to access the absolute (revealed theological truth) for another fictitious claim to access the absolute is philosophically important, since Dionysian frenzy, like religious fervor, may be an intense and powerful experience, but intensity and power need not imply anything about a connection to truth, unless one were to subscribe to a crass pragmatic view of truth.

The musical Dionysian lifeforce of the *Birth of Tragedy* is similar to the Schopenhauerian will or thing-in-itself, since it *seems* to "break through" phenomenal appearances. Schopenhauer refers to music as a superpowerful art that is capable of crashing through the phenomenal modes of expression, describing it as "the immediate language of the will."[15] The strange mood of Schopenhauer appears in the *Birth of Tragedy*, since there is an epistemological limit that is established, but then it is transgressed by a type of poetic intuition. Such a move is bizarre in that the presentation of the noumena yields a type of contradiction if one were to assume that the type of judgment is constitutive. One could defend the claim that Nietzsche's philosophy turns the noumena into a type of poetic literary device, an artistic representation of the unconditioned. No literal explication of the noumena is possible if one is to take it as a purely limiting concept. However, there need not be a contradiction if one is viewing the unconditioned as a work of art. This is similar to what Wurzer refers to as "the aesthetic presentation of the supersensible" in Kant, a peculiar attempt to envision how the world *might be* in Kant's *Critique of Judgment*, a theme further explored in chapter three of our inquiry.[16]

In spite of Nietzsche's apparent lapse into a Schopenhauerian theory of will by the early 1870s in the *Birth of Tragedy*, the FKSP indicates that Nietzsche was not without certain specific reservations about the Schopenhauerian system, even as early as 1867. At the very minimum, this fragment unequivocally indicates that Nietzsche recognized serious theoretical problems with Schopenhauer, especially in terms of the inconsistencies that are present within the enterprise of claiming to know the thing-in-itself. Nietzsche asks where can the dark life force which is named as the will and thing-in-itself appear, if not in the realm of causality and appearance (FKSP 4)?[17]

Even if there are problems with defending the claim that Nietzsche is a Kantian, Nietzsche recognizes Kant as a great reformer in the history of philosophy, without whom Schopenhauer could not have been possible. In fact, Nietzsche's 1867 analysis points out that Schopenhauer's claim to advance beyond Kant is for the most part theoretically bankrupt, a lapse into the imagination that can no longer make a claim to cognition. However, Schopenhauer retains an important value in his brilliant errors, a point to which we must return.

Chapter three of this investigation will explore this important dimension of how to interpret Nietzsche's early encounter with Kant. Nietzsche rejects Kant's claims about ethics, but there still remains a profound debt to Kant in terms of the nature of cognition's root in the phenomenal world, as well as a peculiar debt owed to Kant's exploration of the function of the imagination in the *Critique of Judgment*. Although chapter three of this investigation is devoted to Nietzsche's connection to Kant, some comments about Kant are in order here, given the nature of Schopenhauer's relationship to Kant's philosophy.

In the *Critique of Judgment* Kant refers to the faculty of judgment as a kind of mediator between the sensible and the supersensible: a strange aesthetic presentation of the supersensible is made possible through *reflective judgment*, a technical term which strictly speaking cannot yield cognition, but does have cognitive import for Kant. Nietzsche notes that Schopenhauer lapses into this kind of thinking, a way of picturing the world where the imagination takes on a life of its own. Schopenhauer's claim to know the thing-in-itself is still a product of representation according to Nietzsche's critique of 1867, since the attempt to undertake a theoretical crossing from the phenomenal world to the noumenal (as Schopenhauer himself notes in other parts of his own analysis) yields only "images and words" (FKSP 3).

In spite of the philosophical problems that remain embedded within Schopenhauer's system, Nietzsche still views Schopenhauer as a brilliant misinterpretation of Kant's philosophy. The errors of Schopenhauer prove to be far-reaching, or as Nietzsche puts it, "The errors of great men are venerable because they are more fruitful than the truths of little men,"[18] for a new path was opened for a different type of investigation. It is Schopenhauer who is at the heart of Nietzsche's early biological reflections during this period. Schopenhauer's insistence that the dark will is more fundamental than any type of representation offered a new interpretation and application of Kant's analysis of the limits of human reason. Even if Schopenhauer's thing-in-itself is mixed with phenomenal appearance at times, the shift of philosophical analysis to an underlying reality that animates life proves to be important, as it opened a terrain which philosophy had not adequately explored, a blind spot in the history of philosophical reasoning.

After Kant, God, freedom, and immortality appear as things that are inaccessible to the proofs of reason, yet the importance of recognizing theoretical limits had not been exhausted. Kant introduces the problem of the organism in the *Critique of Judgment*, but it is Schopenhauer who further expounds this line of inquiry.[19] Schopenhauer turns back to bodily experience to underscore the limits of discursive knowledge. One is reminded of Kant's judgments of free beauty in this respect and the significance of judgments of the sublime, both of which suggest a powerlessness of conceptual thought to adequately illuminate certain experiences.[20]

With Schopenhauer we are given a turn back to the body as the ultimate starting point to experience that which we are capable of *knowing* without abstract concepts. This may not be the epistemological sense of knowledge as a justified belief, but something quite different insofar as it purports to yield an immediate awareness. According to this view, a bodily knowledge manifests itself that does not require representation, so knowledge theory not only must be widened to include the traditional conceptual discourse of philosophy, but must also identify immediate bodily experience as a way of knowing. The implications of this seemingly insignificant move are staggering for later developments in Nietzsche's philosophy, Freudian psychology, and Merleau Ponty's bodily phenomenology.

Kant had established important theoretical limits, but Schopenhauer makes a peculiar turn to attempt to transgress the theoretical by appealing to a way of knowing free from discursivity. The attempt to articulate *how* things are known differently and explore the significance of such a turn demanded a re-evaluation of the significance of the body, as well as the significance of the sex drive. By underscoring the limits of discursive knowing, a different way of doing philosophy was initiated, one that challenged the limits of theoretical philosophy. Schopenhauer's view of the knowing subject as a multi-layered contingent structure opened an important terrain for both Nietzsche and Freud. The view of individuation as mere outward appearance of a more fundamental lifeforce certainly undermined any Cartesian view of the self and demanded a renaturalized view of both the individual and the human species. [21]

Although Schopenhauer is heavily influenced by Kant, Schopenhauer's idea of the thing-in-itself is quite different than Kant's. Kant employs the thing-in-itself primarily (but not exclusively) as an epistemological limit, but Schopenhauer stretches the thing-in-itself to encompass many things, at times employing it to point out modes of awareness that are non-discursive.[22] This issue is central to Nietzsche's early analyses that attempt to understand the workings of life, as well as his analysis of the *Birth of Tragedy*.[23]

To articulate other ways of knowing which conceptual edifices misrepresent by virtue of representation qua representation was no small move:

> It is remarkable that in perception, where one man alone is supposed to execute something in an uninterrupted course of action, rational knowledge, application of reason, reflection, may often be a hindrance to him. For example, in billiards-playing, fencing, tuning an instrument, or singing, knowledge of perception must directly guide the activity; passage through reflection makes it uncertain, since it divides the attention and confuses the executant. (WWR I 12)

The claim that discursive structures may misrepresent immediate experiences by virtue of the transference to the abstract was no small move, since it indicated a limit of representational thinking, a blind spot in the history of Western philosophy.[24] Nietzsche focuses on this problem to identify some of the limits that are built into the discourse of traditional Platonism. Insofar as philosophy had to fetishize conceptual knowledge to establish what truth was, something was being left out that could not account for Nietzsche's interpretation of the meaning of music. The origin of music is only one variant of Nietzsche's broader investigations into the limits of conceptual thought and how such limits are connected (or disconnected) with how the world operates.

By the early 1870s Nietzsche posed the question of whether word language was adequate to express all realities. His answer is quite clear: words are important tools of representation, but they are not sufficient to elucidate all experiences. The experiences of music and sex could not be adequately communicated through the discursive structures of word languages alone, and both the history of philology and dominant history of Platonism had very little to offer Nietzsche on these important topics. This is the reason why Nietzsche ventures a poetic expression of the demonic Dionysian in the *Birth of Tragedy*, a path that had been opened up with Schopenhauer's identification of the limits of the principle of sufficient reason. How could most of the history of philosophy in its purported love of wisdom find itself unable to speak sensibly about sex and music? In part it is due to the nature of these phenomena which Schopenhauer ties to the thing-in-itself, but Platonism's impotence on these issues is also due to institutional and psychological factors. Nietzsche found a kindred spirit in Schopenhauer's worldview that recognized how very little substantial knowledge is actually found in universities.

Schopenhauer claims to be the heir to the legacy of Kant, in spite of directing the focus of the Kantian philosophy in a strange new direction. Schopenhauer might be better thought of as a deviant cousin, uninvited to the family reunion of German idealism. The powerful independent character of Schopenhauer captured Nietzsche's attention, for Schopenhauer was committed to finding and expressing truths, even if it turned out his worldview was unpopular, pessimistic, and politically unmarketable.[25] Nietzsche would express his admiration of this side of Schopenhauer in his untimely meditation *Schopenhauer as Educator*. However, as Dan Brazeale and others have pointed out, in that work Nietzsche scarcely refers to any of the specific theoretical details of the Schopenhauerian system. Even if Schopenhauer's theory of the will is epistemologically weak in its self-contradictory nature, Schopenhauer still may be interpreted as valuable both in terms of edification and his ability to look at pain and suffering without fleeing into supernatural accounts of life.

Nietzsche's approach to Schopenhauer during this early phase at times demonstrates an interest in who Schopenhauer was as a person, as indicated by his life experiences and choices that made up his worldview. This *other way* of assessing philosophers reminds one of Diogenes Laertius's style. Diogenes surely had some impact on Nietzsche's analysis, as indicated in chapter one of our inquiry. From one perspective, such an interest in a philosopher's life may be dismissed as unrigorous and missing the point of reading the reflections of philosophers in the first place. However, Schopenhauer's courage to espouse an unpopular and academically unmarketable philosophic worldview captured Nietzsche's attention, since it seemed to offer a type of uncompromising integrity.

Schopenhauer, like Nietzsche, was never widely regarded as a preeminent philosopher of his day. Within their respective fields of philology and philosophy, the reflections of Nietzsche and Schopenhauer would never find wide appeal during their lifetimes. Both held ambivalent attitudes toward academia and both found themselves as outcasts from university life. While Williamowitz and others attacked Nietzsche's creative work in BT by the early 1870s on the grounds of willful distortion, inaccurate philology, and a general lack of sanity by the author, Schopenhauer's reception within the philosophical community was scarcely any better. His amazing erudition could be attacked on the charges that it clearly lapses into contradiction at times, a charge that Nietzsche levels in FKSP. It would be a mistake to underestimate the political dimension that conditioned the reception of Schopenhauer. Beyond his misanthropic tendencies and ad hominem attacks on Fichte, Schelling, and Hegel, a deeper incongruity lurked within academia, one that challenged his integrity as a thinker who strives for truth.[26]

Due to his rejection of physico-theology Schopenhauer was especially vulnerable to the politics of academia. Nietzsche had a deep respect for Schopenhauer's insistence to seek truth and refusal to sell out to the implicit obligations of mainstream philosophers: the obligations to respect state and religious institutions. According to Schopenhauer, virtually no one could get an academic university post without pretending that there were good scientific reasons to believe in the dogmas of religion during the nineteenth century. For Nietzsche, Schopenhauer was a rebel in the cause for truth since he refused to compromise himself as a philosopher. Schopenhauer was not incompetent as a thinker, but rather, he was committed to expressing what he believed to be the truth. Schopenhauer was free from the network of popular metaphysics that demanded political allegiance during his day, even turning his attention to the unpopular question of the nature of the usually unscrutinized class of the priesthood:

> there has never been a lack of persons who have endeavored to create their livelihood out of this need of man's for metaphysics. Therefore in all nations there are monopolists and farmers-general of it, namely the priests. But their vocation had everywhere to be assured to them by their receiving the right to impart their metaphysical dogmas to people at a very early age, before the power of judgment had been roused from its morning slumber, and hence in the earliest childhood; for every dogma well implanted then, however senseless it may be, sticks for all time. If they had to wait until the power of judgment is mature, their privileges could not last. (WWR 2 XVII)

Schopenhauer's view of metaphysics acknowledges the state of human affairs and political environment that helps perpetuate cultural mythologies. Moreover, his larger view of metaphysics as an explanatory principle of the world resonated with Nietzsche's own understanding of the conditions that permit the spread of philosophical ideas:

> By metaphysics I understand all so-called knowledge that goes beyond the possibility of experience and so beyond nature or the given phenomenal appearance of things, in order to give information about that by which, in some sense or other, this experience or nature is conditioned, or in popular language, about that which is hidden behind nature, which makes nature possible. But the great original difference in the powers of understanding, and also their cultivation, which requires much leisure, cause so great a variety among men that, as soon as a nation has extricated itself from the uncultured state, no single metaphysical system can suffice for all. Therefore in the case of civilized nations we generally come across two different kinds of metaphysics, distinguished by the fact that one has its verification and credentials in itself, whereas the other gets its credentials and verification outside itself. Since the first kind of metaphysical system requires reflection, culture, leisure, and judgment for the recognition of its credentials, these systems can be accessible only to an extremely small number of persons; moreover, they can arise and maintain themselves only in the case of an advanced civilization. On the other hand, the metaphysical systems of the second kind are exclusively for the great majority of the people who are not capable of thinking but only believing, and are susceptible not to arguments, but only authority. These systems may therefore be described as popular metaphysics, on the analogy of popular poetry and wisdom, by which is understood proverbs. These systems are known under the name of religions, and are found among all races, with the exception of the most uncivilized of all. As I have said, their evidence is external, and as such, is called revelation, which is authenticated by signs and miracles. Their arguments are mainly threats of eternal, and indeed temporal evils, directed against unbelievers, and even against mere doubters. (WWR 2 XVII)

The view that philosophical inquiry is an expression and function of a peculiar class's dominant values and agenda is an important insight reiterated in the nineteenth century by Schopenhauer, Feuerbach, Marx, and Nietzsche. Such an insight continues to be relevant in the twenty-first century. Marx's view that philosophy may constitute a "surface fluff" is consonant with this view, generating cause to be suspicious of the entire discourse of institutional academic philosophy. Of course Nietzsche does not endorse dialectical materialism and probably would reject the Marxist thesis that all religion, art, and philosophy are conditioned by socioeconomics.[27] Nonetheless, the notion that the proliferation of academic philosophy is directed by non-philosophical power structures and values is a thesis that both Nietzsche and Schopenhauer clearly advocate at times.[28] In other words, power relations are more fundamental than the region that is assumed to be value-free. Philosophy must be paid for, and such decisions are rarely made in a value-free context.

Schopenhauer's recognition that academic philosophy has been policed by theologians remains an unpopular truth. Scarcely does one encounter any courses offered on Schopenhauer in academic philosophy today. Perhaps one could make a case that Schopenhauer is not significant enough in the history of Western thought, but one may still pose the question *why* this is the case: is this the real or only reason?[29] Can such a question be posed from a value-neutral perspective? It is Schopenhauer who points out the systematic nature that had weeded out philosophers who were not amenable to theological representations of the absolute. The systematic use of shame and violence to forbid inquiry into the underlying nature of the body and sex has gone to great lengths. To make persons feel guilty for even questioning the interpretations of the meaning of the body offered by theology appears as nothing less than a systematic brainwashing in Schopenhauer's view. If this view is correct, what a ruling party wants out of a philosophy has directly conditioned which types of philosophers are sponsored by organized religions and political states: what is the purpose of philosophy? If one wants to provide support for religious dogma, then choices of primary thinkers follow. If one wants citizens who obey, then certain thinkers will follow as well, implicitly sanctioned by a power structure that rejects or accepts philosophers according to their tendencies to conform to or resist the preconceived purpose of philosophy.

In a short fragment about the prevailing defects of "Philosophy in the Universities" from 1867 or 1868, Nietzsche reflects on this unpopular Schopenhauerian theme:

> The ruling establishment appoints no one who contradicts religion. Consequently the philosophy of the universities conforms to the religion of the land. An example of this is the Hegelians and their failure. Another one of the ruling establishment's purposes is to appoint philosophy professors who promote the interest of the state. Consequently genuine philosophy goes unrecognized and is silenced to die. (BAW 3 395)

As a mentor who recognized the problems with speaking unpopular truths in the political environment of the university Schopenhauer remained an important thinker for Nietzsche. Yet it would be a mistake to imply Nietzsche simply resorted to an adoption of Schopenhauer's theory of the will. Nietzsche offers a nuanced view of Schopenhauer. Clearly Nietzsche deeply admired elements of Schopenhauer's worldview and character during the early years, but Nietzsche also had important theoretical reservations about Schopenhauer's claim to access the absolute. Schopenhauer's self-sufficient autonomy and strength as a thinker resonated with Nietzsche, but Nietzsche also writes in his 1867 reflections that Schopenhauer's system is riddled with flaws. The publication of the *Birth of Tragedy* in 1872 thus appears as a strange work in Nietzsche's writings, since he seems to repeat some of the same errors that he accuses Schopenhauer of committing in 1867, the most obvious of which is the claim to know something about the nature of the thing-in-itself.

By going back to Nietzsche's earliest and longest philosophic critique of the Schopenhauerian system in 1867, we are furnished with an essential component for accurately assessing Nietzsche's relationship to Schopenhauer. Nietzsche argues that Schopenhauer's attempt to offer a presentation of the unpresentable by claiming to know the thing-in-itself cannot be theoretically sustained. Ultimately, Schopenhauer's discourse on the thing-in-itself attempts to disguise a phenomenal perspective as a grasp of the unconditioned noumenal world. For this reason Nietzsche stresses that any claim to access the thing-in-itself always presupposes a kind of perspective rooted in the world of appearances, for all connections to the thing-in-itself are always falsely inferred from a region of individuation that employs time, space, and causality (FKSP 3).

The move from the phenomenal world to the noumenal world is the sticking point for Nietzsche, a new interpretation of a very old philosophical problem that renders Schopenhauer's work suspect. Ultimately Nietzsche wants to claim in FKSP that Schopenhauer is not able to get to the noumenal without smoke and mirrors, i.e., without using a phenomenal bridge to jump over the gap to the noumenal. The problem is the mediation between the noumenal and phenomenal, a problem that Kant could not resolve in the third critique, and a problem that Schopenhauer is unable to resolve as well.

Nietzsche recognizes Schopenhauer's brilliant errors, and this early work anticipates some of Nietzsche's later warnings against system building as a characteristic of philosophic integrity. Although Schopenhauer has important insights and a provocative worldview, he still does not escape his own contradictory comments on the thing-in-itself. It may be one of the oldest riddles for philosophy—the cleft between the world and how it appears—that still remains a problem for Schopenhauer. In the following selection, Nietzsche offers his most systematic account of Schopenhauer's theory. Although not a polished work, it does furnish us with a clear, conventional epistemological analysis, something that is not often associated with Nietzsche's literary styles. It is doubtful that Nietzsche himself would regard this early critique as significant, given the lack of poetic expression and its fragmentary nature. However, there is a critical philosophical rigor in this piece which makes it very important for discerning precisely what Nietzsche thought in this early phase. Nietzsche specifically attacks Schopenhauer's theory of the will in this early work and identifies some of the serious philosophical problems that plague Schopenhauer's system.

By concluding with passages from the WWR, Nietzsche shows how Schopenhauer himself recognizes at other junctures in the Schopenhauerian system that he should never be able to get beyond the world of time, space, and causality. Schopenhauer's own work can be used to show how he never solves the "oldest riddle of existence." This is a strategy that Nietzsche also employs with Kant, one that follows the thinker's own critique of representation. Nietzsche follows a similar path in 1867 to show that Schopenhauer's claim to find the thing-in-itself is theoretically bankrupt, as Schopenhauer's thing-in-itself ultimately amounts to a disguised type of representation, a phenomenal image that lacks the unconditioned features of the noumenal world.

2. Fragment of a Schopenhauerian Critique (Short Translation)

Schopenhauer tries to explain the world under the assumption that the thing-in-itself can be accessed according to one of its possible forms.

The attempt to do this is unsuccessful.

Schopenhauer held that it was not merely an attempt.

He made his thing-in-itself accessible.

That he himself did not see he was mistaken is to be explained by him not wanting to feel the contradictory darkness in the region where the principle of individuation ceases.

He mistrusted his judgment.

Passages:

The dark drive lying under an apparatus of representation breaks through, revealing itself as world. This drive is not confined under the principle of individuation.

I.

The title page of "the World as Will and Representation" already reveals to us what Schopenhauer intended to announce to humanity through this work. It is the longing question of all metaphysics, as Goethe had announced: "Can nature ultimately be fathomed or not?" This would be boldly answered with a "yes" and with that a new knowledge appears like a holy scripture, wider and further reaching. Thus he had written the redeeming formula for the oldest and most important riddle of the world on the front of this book: the world as will and representation.

Thus that alleged solution recommends a transposition into a metaphorical form of representation in order to comfortably get a hold of the redeeming and explanatory principle to be sought.

The groundless knowledgeless will discloses itself, breaking out from under an apparatus of representation as world.

When we subtract from this proposition what Schopenhauer surpasses from the legacy of Kant, and what he had observed every time in this great man with the most appropriate respect: a word "will" remains his predicate—a difficultly coined, much encompassing word. While others demonstrate a meaningfulness in their own thoughts proceeding beyond Kant, an observer can say of him that he is halted at "that which has been sought under the name of philosophy, and that whose discovery is for this reason regarded by those versed in history as just as impossible as the discovery of the philosophers' stone." [WWR I, preface]

Thereby we are justified in our time in holding that through the old-fashioned ornamented table of categories it was also no small discovery of Kant—appearing as great, as the most far reaching result of his life, although with the characteristic difference that after the termination of "the most difficult task that could ever be undertaken for the purpose of metaphysics," Kant is to step forward as a vigorous breaking through of astonishing natural power and an initiator of a reception—as "The reformer of philosophy," whereas Schopenhauer always knows whom to thank for his alleged discovery, his original presence of mind, and the intuitive power of his intellect.

The errors of great men are more venerable because they are more fruitful than the truths of little men.[30]

Thus when we set to work to lay down the thesis, to analyze the inner conception of the Schopenhauerian system for inspection, there is no thought further from us than to sharply attack Schopenhauer himself with a critique, yet we still reproach his triumphant tone with the demand for a single piece of proof and conclude by throwing out the question with high raised eyebrows: how in the world could a man come to such pretensions from a system so full of holes?

II.

In actuality it cannot be denied that there is an inner conception which is presumed by the Schopenhauerian system, against which we can level the following four objections:

1. The first and most general, against Schopenhauer insofar as he is justified, as he is here not, where it would be necessary to go beyond Kant and to have the concept of a thing-in-itself in view, and speak of it with Überweg, as "only a disguised category."[31]

2. Schopenhauer grants himself justification to follow Kant on the dangerous path, but posits the will in place of the Kantian "x"; only with help from a poetic intuition is this will created, while the attempted logical proof for it is sufficient neither for Schopenhauer, nor for us. (WWR I p. 125. 131).

3. Third, we are forced against preserving the predicates which Schopenhauer attributes to his concept of the will, which, for something absolutely inconceivable, sounds much too determinate and is something won *in opposition* to the world of representation: yet between the thing-in-itself and the appearance not one time does the concept of their opposition have a meaning.

4. Nevertheless, on Schopenhauer's behalf, against all three of these objections one can respond with a three-fold potential possibility: a thing-in-itself can be given, though it is only one based on the region of transcendence, precisely because at any time anything can be hatched out of a brain of a philosopher. This possible thing-in-itself can be the will: because it originates out of the combination of two possibilities, it is plainly still the negative potentiality of the first possibility; in other words, it is already a strong step toward the other pole, which means impossibility. Once again we climb toward this concept always finding a waning possibility for arriving at the predicate of the will itself which Schopenhauer assumes. This is precisely because an opposition between thing-in-itself and appearance is truly unprovable, but it nonetheless can still be thought. Against such a tangle of possibilities each moral thought would be explained. But from this same ethical objection one still can respond that the thinker has no other means than to guess the riddle of how the world stands, i.e., in the hope that in a kinder moment the word lay on his lips, that the key to the riddle would nonetheless be situated for all eyes to observe and be offered in an unread text, that which we call the world. Is this what is signified by the word "will?"—Here is the place where we must make our fourfold objection. The ground of the Schopenhauerian system entangles itself in his hands: in the smallest part it is due to a deliberate tactical unskillfulness of its creator, but it is mostly because the world itself is not comfortable enough to permit becoming fixed in the system, as Schopenhauer had hoped in the first enthusiasm he had as its founder. In his old age he said that the deepest problem of philosophy could not to be resolved through his philosophy. By this, he meant the question that concerns the borders of individuation.

3.

It would be a further specific type of that contradiction which undermines the Schopenhauerian system, pressingly employed. The newborn is a type of this most important and hardly avoidable contradiction, which as it were, while still under the heart of its mother is quietly preparing itself for war and when it is hardly born it does its first deed, in which it kills the mother. They are themselves referred to by the collective borders of individuation and the *first false premise* ([πρωτον Ψ(εῦδος] [proton pseudos]) which the third point touches upon.

"The will as thing-in-itself" Schopenhauer says in WWR [volume 1, chapter 23] "is completely differentiated from its appearance (*Erscheinung*) and fully free from all the forms of appearance into which it first enters when it appears, and which therefore only concern its objectivity, and are foreign to the will itself. Already the most universal form of all representation, that of the object for a subject, does not concern it; still less the forms that are subordinate to this and collectively have their common expression in the principle of sufficient reason; to there it is well known that time and space also belong and consequently also through this principle alone plurality originates and becomes possible. In this last respect, I will call time and space the *principium individuationis*, an expression actually borrowed from the old scholasticism." In this presentation, which we encounter in countless variations in the Schopenhauerian writings, the dictatorial tone surprisingly speaks of a number of negative qualities of the thing-in-itself which lie completely outside of the sphere of knowledge and consequently the claim that this is the most universal form of knowledge of an object is dissonant with his assertion elsewhere that this cannot become an object of knowledge for the knowing subject. Schopenhauer expresses this himself on page 131 [WWR I ch. 22]: "if this thing-in-itself . . . which as such is never an object, (plainly because all of the object is mere appearance, and not the object itself) must be thought of objectively, then we must borrow its name and concept from an object, from something in some way objectively given, and therefore from one of its appearances [WWR I ch. 22]." Schopenhauer thus longs after something which can never be an object yet shall nonetheless become thought of as objective: but on such a path we are only able to long for an apparent objectivity. Insofar as we hang colorful dresses on a thoroughly dark inconceivable "x," they are from a foreign world, as they are taken from the world of appearances. [*Erscheinungswelt*]. The claim is, consequently, that we shall view the thrown on garb as the same as the thing-in-itself: for that is what the proposition means "because it shall become thought of as objective, it must borrow its name and concept from an object." The concept of the "thing-in-itself" would be created furtively from one side "because it thus shall be" and it would be slipped into our hands.

Schopenhauer claims that the borrowed name and concept is the will precisely "because it is the most distinct, the most revealed knowledge, the unmediated illuminated appearance of the thing-in-itself." However, that does not apply for us here: it is more important to realize that the collected predicates of the will are borrowed by us from the realm of appearance (*Erscheinungswelt*). Schopenhauer freely employs this here and then elsewhere presents the attempt to arrive at the sense of this predicate as wholly inconceivable and transcendent; for example WWR II, p. 368: "The unity of those wills in which we have known the being in itself of the world of appearances is a metaphysical one, within which the knowledge itself is transcendent, i.e., it does not pertain to a function of our intellects, and because of this nothing is actually grasped" (WWR I p 134. 132). But we convince ourselves that from the whole Schopenhauerian system, especially to be sure from the first presentation from book one of the WWR that it somehow happens, allowing itself the employment of the unity of the will in a sense which is human (menschliche); yet it still is conceived as completely free from transcendence and fundamentally it is only on the basis of that recurring transcendence, where he presents the gaps in the system to himself as if they were intelligible. This "unity," like the will, as a predicate of the thing-in-itself, is taken from the world of appearances, so its actual core evaporates into the transcendental. It concerns itself with the three predicates of unity, eternity (i.e., timelessness), and freedom (groundlessness), which pass for the thing-in-itself: they are all collected from and bound inseparably with our organization, so it is entirely doubtful whether they have any meaning independent from the sphere of human knowledge. But that he should arrive at the thing-in-itself, because it would lord over its opposite in the world of appearances is proven to us neither by Kant nor Schopenhauer—surely it probably could never be done once, above all by Schopenhauer, because his thing-in-itself, as the will with those three predicates which can never be arrived at and maintained, continues necessarily to be dependent on the world of appearances, i.e., plurality, temporality, and causality is transferred to Schopenhauer's concept of the thing-in-itself.

On the other hand, he maintains this very same position with which we reproach him in full justification in WWR I on page 118 when he says "that from out of the being of things it (the thing-in-itself) will never be arrived at: thus as always one is able to observe that one gets nothing but images and words."

4.

The will appears; how can it appear? Or otherwise put: From where in the faculty of representation does the will appear? Schopenhauer answers with a peculiar turn, in which he indicates that the intellect is the mechanism (μηχανή) of the will (WWR II 315): "But through the evolution of the development of the brain itself there always come more heightened and complicated needs to be led by the corresponding appearances of the will." "The knowing and conscious 'I' would consequently be tertiary in ground, presupposing the existence of an organism, but the organism presupposes the will" ([WWR] II 314). Consequently, Schopenhauer thinks there is an ascending gradation of appearances of the will which continuously climb according to the needs of existence: to satisfy them nature answers with her own ascending gradation of remedies, advancing from an earlier stage of barely illuminated sensation, all the way until the intellect would arrive in its position of utmost clarity. From such a mode of viewing, the phenomenal world still only posits the phenomenal world: this is what happens when we want to retain the Schopenhauerian *termini* on the thing-in-itself. Already before the appearance of the intellect we see the *principium individuationis*, the law of causality in full actuality. The will stirs life in complete haste and in all ways it seeks to appear; it begins in directing the ways of the lowest stages, to rise from the ranks to a certain degree. In this region the Schopenhauerian system is already entirely dissolved into words and images: from the original beginning with the determinations of the thing-in-itself, almost everywhere until the recollection of its original beginning has been lost. And where this advances in between, it once more serves only to show the contradiction achieved in the perfect light of day. Par II. 150: "all of the earth's geological forerunners prior to life would have no consciousness: not in their own, because they have none; not in a foreign, because there would be none. Thus [. . .] there would be none in general; but what then does their having existed signify? It is at bottom a mere hypothesis. Namely, this is because if a consciousness had existed in those primeval times, then such events would have appeared in it; thus far does the regress of phenomena lead us. And so it lay in the very nature of the thing-in-itself to manifest itself in such events." There are, as Schopenhauer himself says on the same page, only "translations into the language of our contemplative intellect."

But we ask after this careful exposition, how was it ever possible for the intellect to originate? The existence of the final step of the appearance of the intellect is certainly just as hypothetical as that of each earlier step, i.e., it would not exist because no consciousness existed. From the next step the intellect should now appear, i.e., from out of a world which does not exist the flower of knowledge shall break through suddenly and unmediated. This should be equivalent to an occurrence in a realm without temporality or spatiality, without the mediation of causality: but what stutters out of such an unworldly world, must be the same—according to the Schopenhauerian proposition—as the thing-in-itself: either it relates to the intellect as a new predicate eternal fastened together with the thing-in-itself, or it can yield no intellect because nothing can become an intellect.

But it does exist: consequently Schopenhauer thinks that it could not be a tool of the phenomenal world, as he wants it to be the thing-in-itself, i.e., the will. The Schopenhauerian thing-in-itself would thus be equivalent to the principle of individuation and be the ground of the necessity: in other words: the world that is present. Schopenhauer wanted to find an equation for the "x" and it is produced from his reckoning that it is = to "x," i.e., he had not found it.

5. Ideas
6. Character
7. Teleology and opposition
8.

The caution with which Schopenhauer proceeds on his path concerning the question of the origin of the intellect is to be noted: we come into the region of this question and hope in silence, that now it will come—that which concealed itself in consciousness from behind the clouds; although it is completely obvious that the intellect in the Schopenhauerian sense is already one manifestation of the individuating principle and presupposes a biased world of the laws of causality. So far as I can see this acknowledgement at one time lay upon his tongue: but he chokes it down in such a strange manner that upon this matter we must go in more closely. WWR II 310 [:] "Now if in the objective comprehension of the intellect we go back as far as we can, we will find that the necessity or the need for knowledge in general arises out of our plurality and the separate existence of beings, from individuation. Thus it is thought to be the same, that it would only be a single essence at hand; thus it permits no knowledge: because nothing is there which would not be differentiated from himself and whose existence is consequently mediated, through knowledge, i.e., he would only be taking up words and images themselves. It would already itself be all in all; consequently there would be nothing to know, in other words, nothing foreign could be apprehended as object. On the other hand, with the plurality of beings, every individual finds itself in a state of isolation from the rest, and from this arises the necessity for knowledge. The nervous system, by means of which the animal individual first becomes conscious of itself, is bounded by a skin; yet in the brain raised to intellect, it crosses this boundary by means of its form of knowledge, causality, and in this way perception arises for it as a consciousness of other things, as a picture or image of beings in space and time, which change in accordance with causality." (BAW 3 352-361). Also appearing in KGW at I 57 [51-5].

3. Recapitulation and Conclusion

Friedrich Lange and Schopenhauer are the most influential nineteenth century thinkers to influence Nietzsche in the mid- to late 1860s. FKSP clearly indicates that Nietzsche did not uncritically adopt a Schopenhauerian theory of the will during his early philosophical development of 1867. The peculiar secondary literature that suggests Nietzsche was completely taken over by Schopenhauer's theory of the will in his early years is oversimplified and misleading. The reason why this misconception continues to persist is due to the explosive Dionysian content of the *Ausser-sich-sein* (literally to be outside of oneself) which Nietzsche puts forth in the *Birth of Tragedy Out of the Spirit of Music*. Nietzsche's view is in fact much more nuanced than a blanket rejection or acceptance of Schopenhauer's theory of the will.

The strange feature of the 1867 FKSP is the chronological priority of it to the 1872 *Birth of Tragedy*. One strategy to make sense of this would be to claim that Nietzsche simply changed his mind about the logical coherence of the Schopenhauerian theory of the will sometime between 1867 and 1871. Nietzsche could have simply changed his mind for a short period of time and decided that the thing-in-itself could be meaningfully disclosed in some sense during that short phase. After all, he entertains conflicting hypotheses in his *Democritea*, at times reversing himself in his notebooks from one page to the next. It does not seem likely that Nietzsche would abort the law of non-contradiction, but he could still try out different types of arguments and styles without feeling obligated to provide a unified view that decisively takes a position one way or the other. However, FKSP is clearly not intended as a hypothetical challenge to Schopenhauer's system, but provides a clear attack on Schopenhauer's inconsistency. Thus it seems completely implausible to suggest that Nietzsche did not really believe during the late 1860s that Schopenhauer's system is deeply flawed.

At first glance the interpretation that holds that Nietzsche simply might have changed his mind about Schopenhauer's system seems to be the most plausible. Thinkers change their minds at times. Although such an interpretation seems sensible, such an interpretation still seems strange, since Nietzsche would have to purge himself of the well-argued positions of FKSP for a short period (between 1867 and 1871), and then return to the position that rejects Schopenhauer's theory of the will shortly after his publication of the *Birth of Tragedy*.

Another possibility of interpretation could be maintained by suggesting that the thing-in-itself that appears in the Dionysian frenzy of the *Birth of Tragedy* could still be interpreted as rooted in the phenomenal world. Thus even this kind of frenzy which has been shared by diverse peoples and places still may be interpreted as being accessed within a spatial and causal world. There is an ambiguity of "timelessness" in the sense that it may either indicate a transcendent region that is completely removed from any causal order, or it may simply indicate a type of common experience that is shared in different historical and cultural epochs.[32] If Nietzsche is employing the thing-in-itself in the latter sense, then there need not be any contradiction in Nietzsche's writings on the thing-in-itself, nor any substantial evolution or revision of his thoughts on the thing-in-itself within his corpus.[33] This seems to be a plausible interpretation, especially since one could maintain that there still are parts of Schopenhauer's theory that Nietzsche found inspiring and insightful. Therefore Nietzsche's embellishing the *Birth of Tragedy* with Schopenhauerian terminology need not suggest a complete acceptance of the Schopenhauerian system. Rather, one could still reject Schopenhauer's larger theoretical claims but still use him as a literary device to attempt to convey the idea that there is something magical, mysterious, and perhaps mystical to the spirit of musical experience which animates the *Birth of Tragedy*—this excess which is beyond measure (*Übermass*) defies a simple linguistic characterization, the dimension that Nietzsche names as the Dionysian.[34]

What possesses Nietzsche to drive himself over the epistemological limit in the *Birth of Tragedy*, to repeat Schopenhauer's errors and write of music as if it were the thing-in-itself? In all likelihood it is Wagner. When Nietzsche first met Wagner in October 1868, he and Nietzsche had an immediate bond through their interest in Schopenhauer's theory of music. Even if Nietzsche never uncritically adopted a Schopenhauerian theory of will, Schopenhauer nonetheless exerted a mammoth impact on Nietzsche's early philosophic development. In an 1868 fragment entitled "the Style of Philosophic Writings," Nietzsche writes:

> What one wants from philosophy is revealed in the judgment of the question of style—whether the purpose is for pure scientific knowledge or to make popular philosophic knowledge, whether it is for instruction or edification. This is the case with the age of Schopenhauer, a healthy pessimism that has an ideal in the background, a powerful masculine seriousness, an aversion to that which is hollow and without substance and an affection for health and simplicity. In opposition to Kant, Schopenhauer is a poet; in opposition to Goethe, he is a philosopher. Compared to Kant, Schopenhauer is naïve and classical. (BAW 4 213)

Nietzsche found a hybrid philosopher poet in Schopenhauer, a mentor who took seriously the limits of representation and stood fast in rejecting popular metaphysics. The "powerful masculine seriousness" which Nietzsche hails in the 1860s seems to be marked by a courageous, self-sufficient philosophic aspiration to identify and criticize the sources of illusion. For good or for bad, Schopenhauer was his own person, politically free to make his own mistakes. His identification of the impotence of academic philosophy to speak truthfully about religious dogma offered Nietzsche an example of what "seriousness" means for a philosopher.

Nietzsche discovered a kindred spirit in the vindictive attacks that Schopenhauer launched against the comfortable illusions of scholastic metaphysics. As was the case with Democritus, Nietzsche found a philosophic integrity and psychological toughness in Schopenhauer, in spite of Schopenhauer's unfounded claim to know the thing-in-itself. Both Schopenhauer and Democritus served as early philosophic mentors for Nietzsche, as both were unpopular for their ideals that sought to dispense with cultural mythologies in the quest for truth. However, Nietzsche's view of Schopenhauer as the embodiment of a healthy character would not last into his later years.

In the 1886 *Critical Glance Backwards*, a later preface to *the Birth of Tragedy*, Nietzsche expresses his regret for fumbling around with Kantian and Schopenhauerian terms to express his youthful musical Dionysian insights from 1871 to 1872. Even as early as 1867, Nietzsche did not agree with Schopenhauer on key theoretical points involving the latter's theory of the will, as demonstrated in FKSP. Although Nietzsche would later reject Schopenhauer as a life-negator in the latter's assessment of tragedy, Schopenhauer remains a key figure for Nietzsche's philosophical development. Schopenhauer's claim that tragedy makes us aware that life is not worthy of our affection takes a different path than Nietzsche's view. Nietzsche's new concept of a life-affirming Dionysian worldview breaks from Schopenhauer's gloomy life-negating pessimism (BT, *Critical Glance Backward* 6).[35]

If one were to look for a constant theme that recurs through most of Nietzsche's philosophy, one could find the concern of "life." Although life itself is a problematic philosophical category, it is clear that Nietzsche returns continuously to this theme, especially in his attempt to evaluate philosophy and history in terms of whether it is life-affirming or life-negating. In the final analysis the later Nietzsche would come to view Schopenhauer's philosophy as embracing a type of life-negation. On this ground, Nietzsche rejects the preoccupation with suffering and misery in which Schopenhauer is caught. Life proves to be the starting point for which philosophy should serve. It is no coincidence that this is also the theme around which Nietzsche drafts an early 1868 dissertation on Kant, the theme of chapter three of our inquiry.

Notes

1. In some secondary literature, Nietzsche's comments about the thing-in-itself have been used as a convenient means to attempt to divide Nietzsche into various phases of epistemological development. Maudemarie Clarke's *Nietzsche on Truth and Philosophy* (Cambridge: Cambridge University Press, 1990) employs this construction by surveying the variant descriptions Nietzsche has of the thing-in-itself, suggesting that Nietzsche had a general early view of the thing-in-itself which would be accessible to human beings. In English scholarship, Claudia Crawford, John Sallis, and David Allison have shown that there are problems with this general approach, given Nietzsche's early specific rejection of the thing-in-itself, a problem to be investigated in this chapter. Although Clarke's interpretation is illuminating in its recognition that there is something different about the *Birth of Tragedy*, it is clear that Nietzsche's earlier work could never support this

interpretation. Quite simply, the *Birth of Tragedy*'s description of an accessible thing-in-itself does not represent a general primitive naive view that Nietzsche holds before 1872. Rather, the strange features of the *Birth of Tragedy* (such as the apparent endorsement of Schopenhauer's theory of the will) are views that Nietzsche specifically rejects in his earlier and later philosophical analyses.

2. Nietzsche did compose work on Schopenhauer that is more lengthy, but only the FKSP provides a specific epistemological analysis of the details of Schopenhauer's system, rather than a critique of cultural significance, integrity, or style.

3. Although drawing very different conclusions, both Kierkegaard and Nietzsche's philosophies were shaped by Kant's groundbreaking claims that the application of human reason is limited to the phenomenal world.

4. In 1888 Nietzsche writes "the will to a system is a will to a lack of integrity" (TI 1 6). Nietzsche's evaluation of system as a vice rather than virtue would seem to target a type of closed foundationalist system.

5. This sense of "original nature" seems to appear cross-culturally in Asian, European, American, and African literature. In this view, there is a desire to return to nature that finds expression in diverse literatures, given the widespread prevalence of an unnatural repressed state in which human beings find themselves.

6. Schopenhauer. *On the Will in Nature*. (New York: St. Martin's Press, 1991), 51.

7. Diogenes Laertius's chapter on Pyrrho caught both Schopenhauer's and Nietzsche's attention by its stress on the conditional nature of individuation, as well the tendency of human beings to overestimate their significance in the context of nature:

> Philo of Athens, a friend of his [Pyrrho], used to say that he was most fond of Democritus, and then of Homer, admiring him and continually saying, "As leaves on trees, such is the life of men." (DL IX 67)

The Homeric comparison of individual human beings to the leaves on trees (*Iliad* Book 6) is helpful to illustrate the conditional status of the individual within the context of the species and nature. Individual leaves have different appearances, but all of them are fated to decompose, just as individual human beings are. However the tree, like the species, lives on.

8. Schopenhauer offers a two-fold analysis to account for awareness in which there is one that is intuitive (the experience of the unmediated will) and one that is discursive (representation). The immediate experiences that Nietzsche describes in terms of the Dionysian have a similarity to Schopenhauer's will, and the image-power of the Apollonian has more than a superficial resemblance to Schopenhauer's description of representation.

9. In section 6 of the "Critical Glance Backward," a later preface of BT, Nietzsche writes: "How I regret now that in those days I still lacked the courage (or immodesty) to permit myself in every way an individual language of my own for such individual views and hazards—and that instead I tried laboriously to express by means of Schopenhauerian and Kantian formulas strange and new valuations which were basically at odds with Kant's and Schopenhauer's spirit and taste!"

10. See Nietzsche's November 9, 1868 letter to Rohde (BW 599).

11. The will "as language" is only understood as a language metaphorically. One of Nietzsche's early interests is how the compositional origin of melody (tone poetry) is

different than the compositional origin of word poetry. Nietzsche's interest in the origin of wordless music searches for the epistemological limit of speech and dialectics to represent music. Nietzsche's focus on the imprisoned Socrates' attempt to compose music in BT touches on this theme, as does Nietzsche's pervasive interest in the significance of the chorus in Aeschylus and Euripides.

12. Germania was founded on July 25, 1860 as a young intellectual group featuring some of Nietzsche's friends such as Gustav Krug and Wilhelm Pinder. Members would present original work to the group that would be subjected to critique and evaluation. Frederick Love points out that "fully half of Nietzsche's own contributions were musical or relevant to music, and some fourteen of these were actual compositions." *Young Nietzsche and the Wagnerian Experience* (New York: AMS Press, 1966). Elizabeth Nietzsche suggests that "Nothing teaches us so much about his intellectual evolution as his essays and reports written for Germania." Elizabeth Nietzsche. *Young Nietzsche*. Translated by Anthony Ludovici (London: W. Heineman, 1912), 97. This claim has merit, as Friedrich Nietzsche's compositions on the demonic power of music and "Pain as a fundamental feature of nature" (a piece of music) all seem to be important embryonic considerations for the *Birth of Tragedy*.

13. *The Birth of Tragedy* is a text about the spirit of music (not words), a theme that enraptured both Nietzsche and Wagner. Although much of Nietzsche's music has been performed and recorded, there is no recoverable record of Nietzsche's improvisational skill on the piano, a limit that could be considered further in the context of his 1868 comment in TSK that "Chance can hit upon the most beautiful melody." In that passage Nietzsche draws an analogy between music and organic life. Neither organic life nor music's origin can be fully understood according to discursive speech structures and notions of a premeditated purposive design. Nietzsche's use of the noumena-phenomena analogy to attempt to explicate the relationship between music and words underscores the limits of the Socratic dialectic.

14. Nietzsche invents things about Dionysian music that are not historically present within the ancient Greek tradition. Silk and Stern point out how rare instrumental music was in Greece until the fourth century. Prior to then, "*mousike* should have meant 'music and poetry.'" M.S. Silk and J.P. Stern. *Nietzsche on Tragedy* (New York: Cambridge University Press, 1981), 137. Williamowitz had already attacked Nietzsche's conflation of the Sileni and the satyr, the latter of which would bear no wisdom, but only partook of lecherous drunkenness. Silk and Stern also observe that "It is the Silenus, not the satyr, who had connections with wisdom, as in the Midas anecdote, which Nietzsche retells with considerable emphasis ([GT] 3). It is on the other hand, the satyrs, and not the Sileni, who have the putative connection with proto-tragedy. Nietzsche's wise, half-animal chorus is his own invention." Silk and Stern. *Nietzsche*, 148.

15. Nietzsche's early 1871 fragment *On Word and Tone* explores the difference between discursive speech and the peculiar "other" power of tones that music affords as a mode of expression (KSA 7 359-69). Rather than describing the will in terms of Schopenhauer's thing-in-itself, *On Word and Tone* describes the will as "the most primitive form of appearance" (*ursprünglichste Erscheinungsform*), yet another example to indicate that Nietzsche did not uncritically adopt Schopenhauer's theory of the will prior to 1872.

16. Wilhelm Wurzer. *Filming and Judgment: Between Heidegger and Adorno* (Atlantic Highlands, New Jersey: Humanities Press, 1990).

17. There still are other ways of interpreting the discourse about the thing-in-itself when it is no longer used as a purely limiting concept. The issue overlaps with what we are to make of speculative metaphysics (in the Kantian sense): if something is presented that exceeds that which is epistemologically knowable, can the presentation still be viewed as a work of art? If so, is it an interesting, life-affirming, or an edifying presentation? Can it be appreciated on aesthetic grounds or for the purposes of psychological critique? Such considerations are similar to but certainly not identical with the fallout that occurred after Rudolph Carnap's attempt to eliminate metaphysics in the twentieth century in *Philosophy and Logical Syntax*. Even if they are epistemologically unjustified, can speculative metaphysics (in the Kantian sense) still remain valuable for the purposes of aesthetic expression or for the purposes of psychological evaluation?

18. This is the only passage of this early fragment that is translated in Walter Kaufmann's *Portable Nietzsche* (Middlesex, England: Viking Penguin, 1982), 30.

19. Kant's critique of teleological judgment specifically deals with how we represent natural organisms to ourselves. They appear within the phenomenal world and seem to be designed. Even though organisms seem to be designed, Nietzsche claims that the apparent design need not have originated from any conscious intention. This problem is explored in chapter three of our inquiry.

20. In the *Critique of Judgment,* Kant connects judgments of the beautiful and the sublime to a subjective universality, thus making aesthetic judgment amenable to his critical enterprise. Such judgments are not rooted in concepts, but have a ground in a peculiar kind of feeling of life (*Lebensgefühl*) in the knowing subject. More obviously, Kant's analysis of teleology raises the question of what concepts have to do with life, an important consideration for biology that will be further explored in chapter three of this investigation.

21. In the sixth meditation Descartes writes:

> Nature teaches me by these sensings of pain, hunger, thirst, and so on, that I am not lodged in my body merely as a pilot in a ship, but so intimately conjoined, and as it were intermingled with it, that with it I form a unitary whole. Were this not the case, I would not sense pain when my body is hurt, being (as I would then be) merely a thinking being, but would apprehend the wound in a purely cognitive manner, just as a sailor apprehends by sight any damage to his ship; and when my body has need of food and drink, I would apprehend this expressly, and not be made aware of it by confused sensings of hunger, thirst, pain, and so on. For these sensings of hunger, thirst, pain, and so on, are in truth merely confused modes of thinking, arising from and dependent on the union and, as it were, the intermingling of mind and body.

Schopenhauer attacks this view of bodily awareness put forth by Descartes in its stress on the idea that the "sensings of hunger, thirst, pain and so on, are in truth merely confused modes of thinking." The type of bodily awareness to which Descartes refers has nothing to do with confused thinking at all, and all attempts to present such bodily processes in terms of a type of thinking obscure the non-discursive nature of bodily activity. Schopenhauer's criticism is valuable because it shows how conceptual reflection can lead to a type of illusion if it is taken beyond the limits of its valid application.

22. In the "A edition" of the *Critique of Pure Reason*, Kant offers the thing-in-itself

as a purely limiting concept. However, the "B edition" offers some reflection on the connection of the thing-in-itself to appearances in the phenomenal world. While it is natural (natural but false) to assume that the thing-in-itself must be something similar to the object of experience that is encountered in the phenomenal world, any such inference would employ the law of causality (the thing-in-itself would "cause" the phenomenal image). According to Kant's own theory this cannot work, since the law of causality is confined to the phenomenal world. The problem here is the connection of appearance to "the way things are." There is no phenomenal bridge that can be invoked to get to the thing-in-itself without violating the general Kantian distinction between noumena and phenomena.

23. Nietzsche explores this theme in his bizarre thesis of the *Birth of Tragedy* that tragedy is ruined due to the moralizing dialectician Socrates. The rumor from Diogenes Laertius (DL II 18) that Socrates begins to direct the composition of Euripides' tragedies is exploited by Nietzsche to offer the thesis that the discursive moralizing dialectics of Socrates subvert the important role of music in Greek tragedy. In other drafts, Nietzsche offers a related variation of this theme when considering modern resuscitative opera and its compositional origin that emerges from abstract ideas and concepts rather than tonal melodies. Nietzsche's nostalgia for a pure Dionysian music that existed independently of words probably amounts to a faith in something that never existed, since the ancient Greeks always conceived of *mousike* in a broad sense that included speech as well as tones. Eduard Hanslick's ideal of absolute music may have colored Nietzsche's view of describing a tonal world completely separate from the order of meaning found within conventional linguistic meaning. The primal experience of music as something wholly different from linguistic meaning seemed to intoxicate Nietzsche's sensibilities, driving him up to and past the epistemological limit in the *Birth of Tragedy*.

24. This may not be a blind spot in Chinese and Japanese philosophical traditions, as evidenced by no-mindedness (*Mushin*) in Japanese martial arts training. Rooted in the Japanese interpretation of Zen, there seems to be more than a superficial resemblance present to Schopenhauer's reflections. Although Schopenhauer's exploration of discursivity as a barrier to certain bodily experiences is quite consonant with such traditions, they probably are not derived from his encounters with Asian philosophy.

25. Schopenhauer's analysis suggests that ideological police have directed the course of academic philosophy to act in subservience to theology, an important insight that has applications well beyond the nineteenth century.

26. Schopenhauer mercilessly condemns many of the most famous philosophers of his day and may not have been the best colleague. In Berlin in 1820 he insisted that his classes be scheduled at the same time as Hegel's classes, so his students' minds could not be ruined by Hegel. Schopenhauer writes, "The public had been forced to see that what is obscure is not always without meaning; what was senseless and without meaning at once took refuge in obscure exposition and language. Fichte was the first to grasp and make vigorous use of this privilege; Schelling at least equaled him in this, and a host of hungry scribblers without intellect or honesty surpassed them both. But the greatest effrontery in serving up sheer nonsense, in scrabbling together senseless and maddening webs of words, such as had previously been heard only in madhouses appeared in Hegel. It became the instrument of the most ponderous and general mystification that has ever existed, with a result that will seem incredible to posterity, and be a lasting monument to German stupidity." (WWR I 429).

27. Marcuse points out that the Marxist view that art is solely a function of the economic substructure may be an oversimplification of the view Marx held. Marx's view of art seems to recognize an important connection to freedom that cannot be accounted for by socioeconomic analysis, even if the so-called Marxist view does not acknowledge this tendency. In the *Aesthetic Dimension*, Marcuse employs Hans-Dietrich Sander's *Marxistische Ideologie und allgemeine Kunsttheorie* (Tübingen: Mohr, 1970) to illustrate this peculiar ambiguity, thus showing that it is possible to defend Marx against the so-called Marxist view.

28. More recently, Carol Gilligan's studies in moral psychology offer much to show how the discourse of truth in moral psychology has been a function and expression of ruling powers that unconsciously validate or reject their own methods and topics of inquiry. The larger and more important issue here is what such details have to do with claims about what is pragmatically true from the perspective of religion, the state, and academia. In the twenty-first century one could make the case that in many philosophical forums there still are religious and state sanctioned imperatives to omit serious discussion about the relationship of money to power, consensual bondage, drugs, masturbation, and the status of Darwinism.

29. Some of Schopenhauer's sexist comments rival those of St. Gerome, and surely this has helped to facilitate his neglect by academic philosophers. In all likelihood, Schopenhauer's rejection of theology has played a larger role in limiting interest in him as a philosopher for academic study.

30. Kaufmann. *Portable Nietzsche*, 30.

31. Lange is probably the source of this connection, as Lange already presented Überweg's objection of the thing-in-itself as a "disguised category," only Überweg's criticism is directed at Kant rather than Schopenhauer. Lange. *Geschichte des Materialismus und Kritik Seiner Bedeutung in Der Gegenwart* (Iserlohn: Von J. Baedeker, 1866), *Zweites Buch*, 267-8.

32. Just as there is an ambiguity in regard to "timelessness," there are also significant ambiguities to the other characteristics associated with the noumena, such as God and freedom. Mohandas Gandhi once remarked that he had never met two persons who had identical conceptions of God. Moreover, Stephen Schwarz points out that there are at least five different senses of freedom. Freedom may refer to political freedoms often associated with liberty, as well as the absence of physical bodily bondage, and bondage within cages. Additionally, there is freedom in the sense of freeing one's mind from prejudice, as well as metaphysical freedom, an issue about which philosophers frequently dispute. Clarifying the precise sense of freedom one has in mind helps eliminate discussions where the disputants talk past one another. Similarly, clarification of the sense of "timelessness" that Nietzsche associates with the Dionysian is important in order to interpret the nature and sense of Nietzsche's exploration of the Dionysian and the thing-in-itself.

33. Such an interpretation is at odds with Maudemarie Clark's suggestion that in *The Birth of Tragedy*, Nietzsche would claim "to know the transcendent character of reality—contradictory, outside of time and space, beyond individuality and plurality—a character clearly quite different from its empirical character." *Nietzsche on Truth and Philosophy* (New York: Cambridge, 1990).

34. *The Birth of Tragedy*'s dictum of interpreting "the world eternally as an aesthetic phenomenon" would thus turn the thing-in-itself into an aesthetic object (as

phenomenon, not noumenon). Such an interpretation of Nietzsche permits one to assess Nietzsche's view of the thing-in-itself as consistent before, during, and after *The Birth of Tragedy*. Such a move would require one to interpret epistemology as a type of aesthetics, a move that is justified and further defended in chapter three based on Nietzsche's analysis of the scope and nature of the aesthetic dimension. However, this still does not explain away Nietzsche's description of the phenomenal world in noumenal terms in BT. Such an issue proves to be a regret of Nietzsche's, as indicated in his later preface.

35. In *Ecce Homo*, Nietzsche would later remark about *The Birth of Tragedy* that "the cadaverous perfume of Schopenhauer only sticks to a few of the formulas" (EH 4).

Chapter Three

The End of Teleology

1. Nietzsche on Kant in 1868

There is no shortage of scholarship that links Nietzsche to the teachings of the *Critique of Pure Reason*, yet few texts connect Nietzsche's early philosophic development (1866-8) with Kant's *Critique of Judgment*.[1] Kant's final critical piece marks a peculiar "end" to the critical enterprise, one that explores reflective judgment, a non-moral, non-cognitive way of representing the world. The *Critique of Judgment* presents a critique of the power of aesthetic judgment, as well as an analysis of teleological reflection, two seemingly diverse types of representation that prove to be important for Nietzsche's early philosophical development. This final chapter about Nietzsche's early philosophy explores how Kant's unique analysis of reflective judgment is influential for Nietzsche's assessment of human reason. Nietzsche employs Kant's view of the human understanding, but he also moves beyond Kant to offer his own synthesis. Such a moment is important in Nietzsche's philosophical development and this early work indicates a consistency with Nietzsche's later claims about the role, nature, and scope of human cognition. In Nietzsche's perspectivism of 1868 there is an important connection to the Kantian insight that the transcendental subject actively structures and organizes the world which it encounters.

In the early months of 1868 Nietzsche began a natural science dissertation to be entitled *Teleology Since Kant*, an investigation that explores the capacity of teleological judgments to illuminate the workings of living organisms.[2] Less than ten years prior Darwin had published *On the Origin of Species*, and Nietzsche was very interested in how human reason attempts to represent and understand the origin and preservation of organic life. Although Nietzsche later criticizes Darwin for his emphasis on survival rather than will to power, Nietzsche recognizes the important significance Darwinism has for humanity.[3] Like the decentering effect Copernicus had on assessing humanity's place in the universe, Darwinism also implied that human beings had no special destiny or purpose within the cosmos, erasing any absolute partitions that may be thought to exist between human beings and the rest of the animal kingdom. In this sense, Nietzsche still shares much with

Darwin, as both seek to understand organic life without invoking supernatural causes.

Like Darwin, Nietzsche embarks on a quest for origins: what made human beings what they are? However, Nietzsche's interests are considerably broader and he further investigates the significance of conceptual reflection: what place does it have in the perceived order of the cosmos? What does life itself have to do with conceptual thought? Darwin's theory offered a plausible account of the origin of life which did not require the postulation of an intelligent designer to account for the patterns in nature.

The idea that life had originated by chance fascinated Nietzsche, and he notes the continuity between the Kant-Laplace hypothesis regarding the origin of the world and the work of Democritus.[4] In both views of the world, there is only a cosmic whirl or vortex that is postulated in the origination of the universe, a blind mechanical force of movement that accounts for the order found in the physical world. However much the Kant-Laplace hypothesis may be similar to the Democritean view of the origin of the world, Kant and Democritus diverge considerably in their evaluation of teleological causes when they are applied to organic life. Ultimately Kant is unable to accept the idea that blind mechanical forces can account for the complex design found in organic life, even though he accepts the notion that blind mechanical forces can account for the order found in the inorganic world.

Chapter one of our inquiry explored Nietzsche's early evaluation of Democritus, the atomistic materialist whose rejection of teleology was censured by Aristotle and a tradition of Western Platonism. More specifically, Democritus appears to have rejected the idea that there was an intelligent designer reflected in organisms. The tendency in the history of atomism to rule out anthropomorphic claims about design impressed Nietzsche. However atoms, like other postulated identities such as a thing-in-itself (either Kant's or Schopenhauer's), turn out to be surface representations for Nietzsche, a claim he assumes in 1868 and later advances in *Beyond Good and Evil* [1885] while admiring Rudjer Boscovich's view of atoms as mere centers of force:

> As for atomistic materialism, it is one of the best refuted theories there are, and in Europe perhaps no one in the learned world is now so unscholarly as to attach serious significance to it, except for convenient household use (as an abbreviation of the means of expression)—thanks chiefly to the Dalmatian Boscovich: he and the Pole Copernicus have been the greatest and most successful opponents of visual evidence so far. For while Copernicus has persuaded us to believe, contrary to all the senses, that the earth does not stand fast, Boscovich has taught us to abjure the belief in the last part of the earth that "stood fast"—the belief in "substance," in "matter," in the earth-residuum and particle atom. It is the greatest triumph over the senses that has been gained on earth so far. (BGE 12)

Nietzsche puts forth this same view of atoms in 1868 in TSK: atoms exist as centers of force within a human network of representation. Thus atoms are not "uncuttables," as their etymology suggests. They are not fundamental particles, but postulates of representation. The theoretical implications of both Copernicus's and Boscovich's work suggest that there is only an illusion of stability in the world, for the earth and the subatomic world are in constant movement and are not immune to the river of Heraclitean flux. Boscovich and Copernicus are important philosophical predecessors for Nietzsche, since they help furnish Nietzsche with a theoretical basis to support the view that the physical world offers only an appearance of stability, rooted in the human perspective.

Kant's own view of himself as initiating a Copernican revolution suggests that the perspective of the knowing subject is crucially important for conditioning the types of knowledge claims humans make. Nietzsche quotes Kant in TSK by stating "Matter is only outer appearance," signifying an overcoming of any type of foundational atomistic materialism. The unity ascribed to atoms is conditional, based on the human viewpoint rooted in appearances, and described by Nietzsche in TSK as "an aesthetic product."

At several junctions, Nietzsche notes that Empedocles had also claimed (even earlier than Democritus) that the parts of the animals were formed by accident. Nietzsche's remark that the Empedoclean insight corresponds to Darwin's theory (BAW 4 54) probably follows from Lange's remarks in the *History of Materialism*, even though other historians had also drawn this connection. From a huge number of chances only the accidents that struck upon a part-whole co-operation survived, but this is not evidence of an intelligent, creative designer in the Empedoclean view, but luck.[5] Chance combined with laws of power and survival offers no deep purpose for life. A purposeless void of the cosmos seems to announce itself through Nietzsche's work, a startling view that searches for intelligent purposes within the universe, only to find an empty abyss. Thus in 1868, Nietzsche offers an interpretation of the significance of life and its meaning in the universe that anticipates his later themes of the death of God and his peculiar version of anthropomorphic perspectivism.

By examining the human perspective as one which creates identities by simplifying and unifying diverse phenomena, Nietzsche formulates a view in 1868 that suggests that human consciousness is much more creative than ordinarily acknowledged. The root of Nietzsche's perspectivism is clearly indebted to Lange's peculiar neo-Kantian acknowledgment of the active role of the human subject. This view is consistent with and anticipates Nietzsche's view of the human intellect in *Truth and Lie in a Nonmoral Sense*.[6] Nietzsche tries to convey a sense in each of these works that the significance of the human intellect is frequently overestimated, a theme that is also clearly present in Voltaire's Pangloss and Nietzsche's caricature of Socrates in the *Birth of Tragedy*.

Teleology Since Kant continues Nietzsche's thematic investigation of the status of teleology in the history of Western philosophy, a theme also addressed and stressed in Lange's *History of Materialism*. Lange's neo-Kantian work clearly persisted in its influence on Nietzsche at least as much as Kant, Schopenhauer, or Democritus. Among these early mentors, Lange is probably the closest to Nietzsche's philosophy. Lange's neo-Kantian views prove to be decisive for Nietzsche's own philosophic orientation and ultimately one may interpret

Nietzsche himself also as a Kantian, but only in a limited and qualified sense, the final theme of this investigation.

Nietzsche was certainly familiar with Hobbes's view of humanity in a state of nature put forth in the *Leviathan* through Lange's work, as evidenced by his citations of the *Leviathan* in TL. Hobbes's modern materialism attempted to view nature without nostalgia for metaphysical ideals, and such a view resonated with Nietzsche's view of nature. Both Hobbes and Nietzsche claim that good and evil are neither simple nor absolute, and these concepts, as well as the meaning of justice, are products of human convention. More important for Nietzsche's work on Kant, however, is the Hobbesian view of nature as a violent place, a war of all against all which does not suggest any type of intelligent moral designer. This view is important for Nietzsche's philosophy, since it suggests that moral interpretations of nature are fictitious human viewpoints. Moreover, if moral interpretations are merely an all-too-human way of looking at things, there still is a philosophical need to explain why such widespread illusions have seemed so plausible to so many different peoples in diverse times and places. To make sense of how this is possible, Nietzsche needed to furnish an account of the deeply creative nature of the human intellect. Evaluating the significance of moral judgments within nature is only part of Nietzsche's larger project of investigating the extent to which anthropomorphic judgments *color* our grasp of the world.

In order to assess the power which allows humans to introduce unity and stability into the world one must move beyond mere psychological analysis, as psychological factors still do not fully account for the widespread view that there is an intelligence within nature. In short, why does it really *seem* like there is order, purpose, unity, and design in the world? Nietzsche's early work on Kant focuses on the origin of such ideas (order, purpose, unity, design) by offering an account of their origin as a distinctly *human* product of organization.

Teleology Since Kant gives us a glimpse of Nietzsche's early nuanced view of Kant. During this phase Nietzsche is critical of Kant, but still appreciates Kant's deep contribution to the history of philosophy. This phase provides a different view than in his later writings where he calls Kant "a scarecrow" (*Vogelscheuche*) (KSA 11 263), "an underhanded Christian" (TI III 6), and "an idiot" (A 11). At other junctures Nietzsche trashes Kant for lacking personality, suggesting that an account of his life could only amount to a "biography of a head" (D 481).[7] Nietzsche's later claim that "a will to system is a will to a lack of integrity" could be aimed at the *Critique of Judgment*, since Kant announces his intention to complete his critical enterprise in the preface of the *Critique of Judgment*, a completion that consists of a critique of *the power of primal parting*. However, one could interpret the *Critique of Judgment* very differently: although it is supposed to complete Kant's system by reconciling the seemingly divergent realms of nature and freedom that he had advanced from the *Critique of Pure Reason* and the *Critique of Practical Reason*, Kant does not deliver a closed system. Kant presents a type of non-cognitive and non-moral judgment that imagines how the gap between phenomenal nature and noumenal freedom could be bridged, an odyssey into *how* the imagination works to present a unified view of the world to the transcendental subject.

Kant's portrait of reflective judgment elucidates a special type of technical judgment within his system that is capable of unifying diverse experiences. The power of reflective judgment reveals a deep capacity of the imagination to strive for unity in its representations. Kant observes that such a power allows human beings to represent different laws and subsume them under more fundamental laws. Reflective judgment helps to make experience coherent by unifying different types of sense data, laws, and observations. It is a condition that makes possible new hypotheses about the world. Insofar as reflective judgment works to assist the subject in the postulation and creation of identities and unities, it demonstrates a deep affinity with Nietzsche's wider view of the human cognitive faculties.

Reflective judgment reveals different capacities of and for the imagination, capacities that Kant had not explored in the *Critique of Pure Reason.* The human understanding requires the imagination in its representational capacities to make cognition possible in the *Critique of Pure Reason,* a move that elevates the imagination's significance for making claims about knowledge within the history of philosophy. Kant actually considers the imagination to be the deepest of the cognitive faculties.[8] While Nietzsche's philosophy would appear to be opposed to the spirit of Kant's critical enterprise in many respects, Nietzsche's perspectivism may also be interpreted as a continuation and augmentation of Kant, especially in regard to the active, unifying role of the transcendental subject and the ideality of space and time.

Any attempt to consider Nietzsche's early philosophy as a response to the third critique must also consider the first critique, since the *Critique of Judgment* must be read from the perspective of the larger critical enterprise to fully understand and appreciate the overall significance and intentions of Kant's final piece of the critical enterprise. There is much scholarship devoted to Nietzsche's ambivalent relationship to the *Critique of Pure Reason.* During his earlier years he has a tendency to view Kant in a favorable light, recognizing Kant as a great reformer, indeed the great philosophic reformer in modernity. There is a greater tendency to ridicule Kant's personality, view of the aesthetic dimension, and ethical theory in Nietzsche's later work, and there are comparatively few references to the *Critique of Judgment* in the texts which Nietzsche published himself.[9] However, Nietzsche was deeply influenced by Kant, since he retains an important aspect of the critical enterprise, one that limits the knowledge claims of the transcendental subject to the realm of appearance. Kant's exploration of the imagination in the third critique offered still much more on the imagination than the first critique, an issue that consumes Nietzsche's attention in 1868.

From the point of view of knowledge theory, Kant's division of the world into noumena and phenomena in the first critique assumes some strictures on the knowing subject by limiting human beings to the world of appearances. Nietzsche, like every other serious philosophical thinker of the nineteenth century, could not ignore the *Critique of Pure Reason.*[10] Kant's investigation into the scope and nature of the objects of knowledge opened up an important path for Schopenhauer and Lange that was continued by Nietzsche.

In the *Critique of Pure Reason* Kant advances the groundbreaking claim that human beings are limited to a world of finite sense intuitions to access the world. The knowing subject qua subject conditions its own experience via the forms of sensibility: the knowing subject experiences phenomena in space and time, but space and time are merely the conditions that are "within" the subject. Kant's view suggests that we only experience the phenomenal world, so human beings encounter the world only as it appears. Things as they are in themselves cannot be known. Human beings thus never cognize the external world as it actually is, since the parts of the world which are investigated must conform to the structure of the human mind which makes experience possible.

In the Kantian view, "speculative metaphysics" attempt to go beyond the conditions for the possibility of experience. Kant explains why there have been interminable battlefields of metaphysics that have raged for centuries with little or no progress in the *Prolegomena to Any Future Metaphysics.* Unlike the natural sciences (such as Newtonian physics) which rightfully deserved to be called scientific, metaphysics had made little or no progress in over two millennia, since it had not properly critiqued the powers of the human understanding. Kant demands that claims to knowledge (of the "external" world) must be grounded in some kind of sense intuitions. In large measure, speculative metaphysical disputes about the existence of God, freedom, and immortality are incapable of being resolved because such matters have no possible corresponding sense intuition for the transcendental subject. Such a situation suggests that there are no experiences that could count as evidence to confirm or disconfirm the existence of the noumena.

Nietzsche amplifies Kant's project by suggesting that there are problems with any system of metaphysics, moving beyond both Platonic idealism and any type of foundational atomistic materialism. The subject's "optics" which alone make experience possible are always mediated by the workings of a fictioning power: the knowing subject does not get the world as it is, but only as it appears through the transcendental subject's forms of intuition. In this sense, the apparatus of the human understanding relies on identities that only appear in the field of representation. Nietzsche offers a distinctly aesthetic view of the world, since the human understanding takes on a similarity to that of a painter's easel: the creative process that is usually associated with and confined to the artistic imagination bears a striking similarity to the organizational processes of the transcendental subject. Nietzsche's later comments about "the optics" of the knowing subject are in some ways similar to the Kantian categories and forms of sensibility.

As in English, there is in German an ambiguity with anything that appears. Joan Stambaugh has pointed out that "appearance" can either have reference to illusion,[11] i.e., the glass appeared to be empty, but it was filled with clear liquid (appearance as illusion), or "appearance" may indicate actuality, i.e., the fox appeared in the clearing (appearance as disclosure, without reference to dissimulation). Nietzsche interprets the world of phenomenal appearance to be a type of illusion in this former sense, a problem that still lingers in the final analysis for both Schopenhauer and Kant.[12] Nietzsche's view that "truth is the relationship of various errors to one another" is indebted to Kant's view that the knowing subject only gets appearances.[13] Thus "appearance" can be understood as a type of perspectival illusion here, rather than the sense of "appearance" that is free from dissimulation.

Nietzsche's relegation of conceptual thought to a surface play in *Teleology Since Kant* expounds Kant's view that there are aspects of the world that cannot be adequately represented. Kant's claim that we can never have knowledge of God, freedom, immortality, the infinite, or the thing-in-itself proves to be important for Nietzsche's interests. Kant is no atheist of course and he himself believes that his critical enterprise is laying the framework to mediate between genuine and bogus claims to knowledge. Nonetheless, his critique of speculative theology is ideologically friendly to atheism, since it implies that speculative theology is a kind of fiction.[14] If not fiction, it nonetheless is incapable of making any direct claims to knowledge about the world, an implication of the critical project that Schopenhauer recognizes as perhaps the most significant contribution of Kant to modern metaphysics.

Nietzsche's view that the history of philosophy is often fueled by a desire for rationalizing the ethical dimension obviously includes Kant. Nietzsche's relegation of moral theorizing to the realm of human illusion obviously breaks from Kant's conception of the postulates of practical reason.[15] From one perspective, Nietzsche's denial of noumenal freedom would appear to rule out the possibility of interpreting him as a second or third generation Kantian. On the other hand, it is possible to view the noumena as Nietzsche appeared to: as a plaything of aesthetic representation, authored by a fictive representing power.

Nietzsche follows much of the Kantian interpretation of the limits of human reason, but varies considerably on the nature of ethics. There does appear to be an ethics in Nietzsche at times, one that demands intellectual integrity [*Redlichkeit*], as well as an injunction to remain faithful to the earth.[16] I think that there is also a coherent and important message to Nietzsche's philosophy, one that presupposes the slippery (but not empty) philosophic categories of "health" and "life-affirmation." However, it certainly is worth noticing that there have been many illusions perpetrated in the name of such concepts, especially with regard to the concept of "health."

The term "health" is of course by no means unambiguous, since it may either refer to a relationship that is biologically immanent within an organism, or it may refer to a mode of representation that is more obviously value-laden.[17] This problem is compounded further when the concept of health is extended to a culture, a polity, or metaphysical and moral systems. Although there are problems with identifying a consistent unified sense of what constitutes "health" within the Nietzschean corpus, it is important for Nietzsche's understanding of sickness. Whether rooted in power or a notion of balance, Nietzsche's concept of health at times suggests a disdain for metaphysics. There clearly is an important connection of the concept health that overlaps with moral evaluation in Nietzsche's thought, an important and difficult topic which exceeds the margins of this investigation.[18]

Even if there is a very different interpretation of the ethical dimension present in Kant than in Nietzsche, there clearly are some aspects of Kant's analysis of the limits of human reason that Nietzsche accepts. More problematic than their thoughts on the ethical dimension or knowledge theory is the two thinkers' assessment of the nature and scope of aesthetic judgment. The matter is exacerbated by Nietzsche's seemingly cryptic comments that aesthetic forces are more fundamental than the knowing subject.[19] Such a view suggests that the aesthetic dimension has made possible the knowing subject, rather than the usual conventional view that maintains the knowing subject makes possible the aesthetic dimension. Such assertions need not be thought of as incoherent if one considers Nietzsche's reference to his own philosophy as "inverted Platonism" by the early 1870s, a suggestion that art is the most real, rather than the Platonic view which holds art to be the least real.

Although clearly at odds with Kant's view of his own work, it is indeed possible to interpret Nietzsche as offering a philosophic re-interpretation of the Kantian project, envisioning the constellation between the first and third critiques as an aesthetic formation.[20] One could take Nietzsche's claim that "the world is justified eternally only as aesthetic phenomenon" (BT 5, 24) as an invitation to interpret the third critique as an aesthetic continuation of the first critique of the noumena-phenomena split. Although this certainly would be contrary to Kant's view of aesthetic judgment, the noumena-phenomena split could be interpreted as an aesthetic transformation in which the judgments of the sublime and the beautiful correspond to the noumena and phenomena. The noumena finds expression in the attempt to "take in" the infinite as Dionysus, while the phenomenal world finds expression in the restful contemplation of the beautiful, the shining dream world of Apollo.[21] Even though cognition of the infinite is ruled out of bounds by Kant, there still is the attempt to present it aesthetically in the analytic of the sublime. In spite of the differences between aesthetic and cognitive judgments, there is an underlying unity of the difference between the finite and the infinite that Nietzsche explores. Following such a path, one could interpret epistemology as a type of aesthetics. However, such a speculative view considerably broadens the domain of what is traditionally thought of as aesthetics in the modern world.

If this line of comparison of Nietzsche's view of aesthetic judgment were extended, a synthesis of the noumenal-sublime-Dionysian and the phenomenal-beautiful-Apollonian could be envisioned. Many thinkers would reject and have rejected such an interpretation of Nietzsche as far-fetched, suggesting that Nietzsche could never have produced such a response, as if he lacked the tools for reinterpreting Kant's first and third critiques directly. Heidegger, for example, claims that Nietzsche has a thoroughly colored view on Kant's teaching of the beautiful because Nietzsche is alleged to have only read Schopenhauer's interpretation of Kant.[22]

In spite of the diverse interpretations of Nietzsche that abound in secondary literature that have claimed that Nietzsche did not read Kant directly, there is no good reason to believe Nietzsche did not carefully read Kant, as this is evidenced by his page citations of the 1838 Rosenkranz edition of Kant. More importantly, even if Kant and Nietzsche exhibit very different types of philosophy in terms of their thoughts on epistemology, ethics, and aesthetics, there still are similarities present in these two thinkers. Kant is not a skeptic, but his view of cognition clearly scales back what the subject may know by delimiting it to the phenomenal world. Kant's stress on the limits of human reason directly contributes to a philosophic climate of the nineteenth century that helped create Nietzsche's distinctly modern aesthetic perspectivism. Nietzsche's view that the history of Platonism has overestimated the significance of human reason's ability to fathom the nature of reality is directly indebted to Kant's recognition of the limits of reason. There have been many different ideas associated with what the term "Nietzschean" means, but comparatively few thinkers outside of the French interpretations of Nietzsche have taken this tendency seriously until fairly recently.

I do not believe the reason for ignoring Nietzsche's deep skepticism about truth and human reason is politically motivated in nature.[23] However, very few propagandists enlist skeptical writings to advance their causes. In skepticism, there is usually something decidedly unuseful for advancing most political agendas, as Hume and others have noticed.[24] The dissemination of philosophic texts is often political in nature, and this dimension has conditioned the investigation of Nietzsche's surviving published and unpublished works. Nietzsche's extreme skepticism about human knowledge and the relativity of truth has frequently been viewed as an afterthought, rather than occupying the core of his philosophy. However, this issue was always important for Nietzsche, since it conditions the question: what can metaphysics tell us about the nature of life?[25]

In spite of their apparent diversity, Kant and Nietzsche both found themselves involved in the business of critiquing epistemological limits, and Kant's claim that speculative metaphysics can never be knowledge resonated deeply with Nietzsche. Kant's *Critique of Judgment* offered an important final chapter of the Kantian project for Nietzsche, since it focused on art and forms of life, two themes that are at the heart of Nietzsche's enduring philosophic interests. Clearly Nietzsche was very interested in how the concepts of art and life posed special challenges to the investigation of epistemological limits.

The notes which constitute "*Teleologie Seit Kant*" indicate Nietzsche's early desire to establish how and why humans know what they claim to know. Much of it consists of quotes from Kant piled upon other quotes which respond to Kant's philosophy, thus imbuing it with a fragmentary appearance which justified its relegation to "Juvenilia" in the Musarion edition of Nietzsche's works in the early twentieth century. Beyond their "borrowing" very heavily from Kuno Fischer's fifth volume of *Geschichte der neuern Philosophie*, these philosophic notes reveal that in 1868 Nietzsche was already critiquing the limitations of the intellect as an anthropomorphic fictioning power. Nietzsche's theory of interpretation as a network of perspectival errors is already explicitly contained in this work. Nietzsche's dissertation draft attests to the deep influence of Lange's brilliant survey of materialism.

Before examining Nietzsche's 1868 analysis of Kant in detail, it will be helpful to consider some of Nietzsche's other comments directed at the limitations of representation in other parts of his corpus. By doing so it will become apparent that *Teleology Since Kant* indicates a continuity with Nietzsche's later peculiar version of aesthetic perspectivism. By examining Nietzsche's comments on Kant from the 1870s and 1880s, along with his other comments that are directed at the role the transcendental subject plays in claims to knowledge, Nietzsche's theory of the human understanding will be clarified and explicated. In spite of what are often viewed as contradictory comments directed toward epistemology, there still is a consistency that pervades Nietzsche's corpus, one that rejects the existence of absolute and supernatural truth, and *Teleology Since Kant* indicates that he had arrived at this general worldview in 1868. As Nietzsche points out, frequently contradictions that are found in texts by critics are actually created by the critic's interpretation of the text, rather than being present in the work itself.[26]

Nietzsche writes "every identity is imaginary" in the 1860s. However, this reference initially occurs within the context of describing the identity of Homer and the origin of this peculiar sense of identity.[27] Nietzsche transfers this textual deconstruction into a wider arena that critiques the limits of human reason, a movement from literature to metaphysics. This broad critique of identity attacks the ground of any stable metaphysical identities by suggesting that the identities are only a product of human representation. The view that humans continually create contingent identities that are not acknowledged to be contingent in nature is one that takes seriously the active form-giving nature of the human understanding.[28]

This deeper view of the subject as a creative author of identities within the field of representation is expounded in some detail in *Truth and Lie in an Nonmoral Sense* (1872), where Nietzsche suggests that we (human beings) are artists more than we recognize. By stressing the dissonance between the real and the ideal, Nietzsche shows how the Platonic form (as class concept, universal, and ideal essence) cannot adequately account for individual "reals."[29] The movement between (ideal and real)/(abstract and concrete) is one of the senses of the powers of judgment which Kant first elucidates in *Critique of Pure Reason* as the faculty of subsuming under rules (KRV a133). Kant does not conceive this as an aesthetic movement,[30] but there is a sense in which Nietzsche appears to interpret it as such, conditionally collapsing the distinction between various kinds of judgments.

In *Truth and Lie in a Non-moral Sense*, Nietzsche suggests that there is a sense in which determinative judgments are also aesthetic (in the "artistic" sense), highlighting that "everyday consciousness" organizes and manipulates phenomena: we do not think about whether "the leaf" is subsumed under the abstract "universal (*Urform*) of leaf" (TL 1).[31] Throughout Nietzsche's published and unpublished work he suggests that the world seems like a rational place because humans have imposed order on the dynamic perpetual flux. Human beings unintentionally organize the world which they experience, since a process of organization is presupposed within the knowing subject to make experience possible. Moreover, human beings also organize the world by the very use of class concepts. This view is at the core of Nietzsche's aesthetic perspectivism, an interpretation that acknowledges a radical limitation of the human view:

> To be completely truthful—a magnificent heroic wish of humanity, in a lying nature! It is only possible in a very relative sense! That is tragic. That is the

> tragic problem of Kant! Now art receives a completely new dignity, against which the sciences are degraded. Truthfulness of art: now she alone is honest. So we find ourselves on a strange roundabout path, again back (like the Greeks) to the natural condition. It has proved itself impossible to build a culture according to knowledge (*The Philosopher's Book* KSA 7 19 [104]). [1872-3]

After Kant had barred epistemological claims to the absolute in the *Critique of Pure Reason*, the tenability of theoretical proofs for the existence of God had been relegated to the realm of speculative metaphysics.[32] Not everyone accepted this tremendously disruptive move, but it is clear that the transcendent absolute had become theoretically problematic, much more so during this time than in any prior movement in the history of Western philosophy. While Kant still allows for an ethicotheology from which Nietzsche sharply breaks (KU 86), Kant had revoked the plausibility of retaining determinative claims to knowledge by any physical teleology in the *Critique of Judgment*. It is this "end" of the critical enterprise which commanded the philosophic reflection of Nietzsche in the months of 1868.

It should be no surprise that the young Nietzsche took an interest in the *Critique of Judgment*, given Kant's attempt to understand the judgments rational beings make about the aesthetic dimension and organic life in the final critique. Kant offers a peculiar work that connects aesthetics and teleology together, and the different mood of this text did not go unnoticed by Hegel, Schelling, or Schopenhauer.[33] The relation between art and nature in the *Critique of Judgment* is non-oppositional, disclosing that both nature and the "artwork" are on both sides of the knowing subject.

At the beginning of the *Critique of Judgment*, Kant describes the feeling of life (*Lebensgefühl*) at the basis of aesthetic judgment. This offers a new direction for his critical philosophy, since the pleasure within the knowing subject is the starting place for investigating the aesthetic dimension. Kant explicitly asserts that aesthetic judgment is impaired if one judges aesthetic "objects" merely in terms of concepts (KU 8). The enlivening of the knowing subject that is connected to the feeling of life suggests a limit of what can be expected from conceptual thought. Kant's analysis of the feeling of life is not concerned with sensual gratification or Dionysian excess, but it still serves to help map out the limits of purely conceptual thought to explicate the aesthetic dimension.

Kant's feuding with the "*Sturm und Drang*" movement has been more than adequately documented. He seems to have instinctively cringed away from the thoughts on art on which he found himself writing, perhaps recognizing that he was opening up a "dark side" in his attempts to fully explicate the limits of reason. The "Nature" which acts *through* the genius by giving the rule to art is certainly a mood which is quite foreign to the first two critiques. Both Kant's conception of the genius and his analytic of the sublime provide a very different sense of nature than that found in the *Critique of Pure Reason*. "Original nature" at times breaks through the transcendental subject in the *Critique of Judgment*, bearing a similarity to Nietzsche's concept of the Dionysian. After doing so much to combat speculative metaphysics in the first critique, Kant does not want to support art as a means to launch a cult of fanaticism.[34]

Sometime after publishing the *Critique of Pure Reason* in the 1780s Kant decided that aesthetic judgments were also amenable to transcendental argument.[35] The violent shuddering of the subject and "the violence done to the inner sense" by which time is annihilated (KU 27 258-9) in the analytic of the sublime appear quite foreign to the mood of the transcendental subject that Kant presents in the first critique. Kant's peculiar presentation of the feeling of the sublime as a universally shared feeling is a strange point in the Kantian project, in that it has the special focus on *how* the knowing subject feels it limits in the context of the raw power of "original nature."

Kant considers the status of idols in the *Critique of Judgment* and why such representations of the infinite were barred in the very old Jewish tradition. After Kant ruled out the possibility of having knowledge of the infinite in the *Critique of Pure Reason*, the question still remained as to what the individual transcendental subject's relationship to the infinite could be. Even if there is no theoretical knowledge of the infinite in Kant's critical philosophy, there still are attempts to represent the infinite. Kant's analysis of the sublime offers an attempt to represent the infinite aesthetically. When the might of nature is encountered, there is an outpouring of vital forces and shuddering that occurs within the subject. Although Kant does not want to suggest that the shuddering triggered by the power of nature reveals something about the subject's conditional nature, Kant does think past the region of transcendental subjectivity with his explanation of the genius in this work (KU 46). In Kant's analysis of the genius in the *Critique of Judgment*, original nature acts through the subject to give the rule to art. Thus there is a similarity in Kant's attempt to offer an explanation of the creative power of the genius to Nietzsche's view of the Dionysian in that the partition between the subject and "outside world" is not absolute.

The power of creativity in Kant's analysis of the genius is described as something original and literally seems to come from another world. It is as if the transcendental subject is caught in the crossfire, "in the way" of original nature's own direction of the work of art (KU 46). There is a kinship of Kant's exploration of the origin of art to Nietzsche's view of the demonic and its relation to music from the early 1860s. Nietzsche's early thoughts on the demonic clearly are continuous with and probably evolve into his later description of the Dionysian. In both cases, there is an uncanny eruption of nature, one that is channeled through the artist. Both Nietzsche and Kant try to explain the original power of aesthetic forces that seem to burst forth without mediation from a human artist, and Nietzsche frames his description in the *Birth of Tragedy* with Schopenhauer's language of the *principium individuationis*, the individuating principle.[36] The lapsing of the principle of individuation allows one to peer beneath human conventions to recognize how one stands in relationship to the natural world. Although Kant does not draw this conclusion, this occurrence that makes the individual feel small is important, since it can help trigger a type of insight into the precarious conditional nature of the self, a unity that probably lacks any permanence.

Although Kant does not think the infinite can be an object of knowledge, he nonetheless assumes that there is a type of design that is present in organic nature that simply cannot be accounted for by randomness. In his attempt to understand nature, he feels compelled to assume that there is an order in the cosmos that is also reflected in organic life. "Nature's technic" reveals a peculiar ambiguity which begins to make it difficult to articulate where artworks begin. Kant's attempt to understand life itself seems to imply that there can be artwork that is more fundamental than the transcendental subject:

> It is true that when we judge certain objects of nature, above all animate ones, such as a human being or a horse, we do commonly also take into account their objective purposiveness in order to judge their beauty. But then, by the same token, the judgment is no longer purely aesthetic, no longer a mere judgment of taste. We then judge nature no longer as it appears as art, but insofar as it actually is art (though superhuman art [*übermenschliche Künst*]). (KU 48 311)

Although appearing to be two different types of judgment, the connection of teleology to aesthetic judgment may be spawned from the traditional view of God as divine artisan; the question remains, "Who or what is the author of superhuman art?" Do the creations of nature point back to a conscious designer, or is this Kant merely wandering over the epistemological limit into a fantasizing of the supersensible? Even if there were forerunners such as Empedocles and Democritus who suggested that life was generated by chance, the historical responses to such questions have been investigated and considered differently before and after Darwin.

Kant's suspension of any type of determinative judgment within the realm of physical teleologies opened up new possibilities in that he showed that physical teleologies are problematic within the domain of natural science. Even though Kant loses sight of this limitation, he still undermines the possibility that physical teleologies can offer direct knowledge of the world. Nietzsche attempts to out-Kant Kant by suggesting that the aesthetic-teleological presentation of a superhuman artist makes a leap to a realm into which a finite being has no access. Reflective judgment is only able to envision how the world *might* be, but there are alternative possibilities that may be presented which do not require any assumptions about a supernatural director.

Nietzsche's early work in TSK shows how Darwinism had proceeded further in Nietzsche's day to suggest that life is much older than prior generations had imagined. By ruling out a master designer, Nietzsche attempts to show that there is more evidence that life was generated by accident. This view that life was much older than the traditional accounts offered by Western theology is humbling, since it implies that there is no special destiny for human beings in the universe. Since Nietzsche, fossil records have become much more extensive, and developments in carbon dating, radiometric dating, and potassium-argon dating offer still more powerful evidence that one need not presuppose the recent master artisan that most modern Western theologies had assumed in their attempts to account for the creation of organic life.[37]

If one accepts the proposition that there is no logical need to assume a conscious designer in order to account for the design presented by the existence of organic life, one must still account for why and how it is possible for such representations to occur. In short, why does it seem like there is a conscious design present in complex organisms or even in relatively simple life forms? According to Nietzsche, this is due to a tendency to view the world according to a distinctly human perspective, as the idea of intelligent design exists according to a human analogy. Such an explanation is virtually identical to Friedrich Lange's.

Nietzsche identifies Kant's larger expanded presentation of teleological order as an aesthetic portrait of the cosmos. The *Critique of Judgment*'s portrait of the cognitive faculties suggests that reflective judgment creates unity and coherency, a means to construct a fabric of ideality that helps to explain the world. It is this basic feature of reflective judgment that resembles Nietzsche's general view of the human understanding, as both suggest a process that unifies complex and diverse phenomena, a type of synthesis that allows the subject to organize the world. Kant's exploration of reflective judgment illustrates the deep capacity of the subject which strives to make the world seem coherent, an important feature of representation that reveals a deep creative power of the subject that is usually unacknowledged. One could even argue that this basic process of making the world coherent is a form of the will to power, since the unification and simplification permits a type of mastery over the brute diversity of nature. This is one of the ways in which Nietzsche argues that some errors are valuable for empowerment, in the sense that simplification is a type of perspectival error. Connecting this feature to a larger view of the will to power lies beyond the margins of this investigation.

The power of the transcendental subject to organize phenomena is an important dimension for assessing the creative role that human awareness presupposes. Although the world may appear to be organized at times, Nietzsche's work in 1868 attempts to show that such inferences are based upon the failure to recognize that the organizing features emerge from the network of representation. Nonetheless, there still remains the power of the organism, something that still seems mysterious in Nietzsche's view in that it maintains itself without thought. Both Kant and Nietzsche view the inner workings of the organism as something elusive, challenging and exceeding the limits of conceptual reflection. The unconscious process by which organisms maintain themselves provides a mystery that is not found within the inorganic. Indeed, organic life is different than plate tectonics: life forms seem to be organized in complex ways which are not exhibited by the inorganic in their ability to maintain and reproduce themselves.[38]

In TSK Nietzsche responds to Kant's postulation of a "natural purpose," Kant's attempt to represent the inner workings of the organism via reflective judgment. Nietzsche ultimately wants to point out that it is possible to explain "natural purposes" as originating from accident. Kant claims that we are able to consider something as a "natural purpose" if it is both cause and effect of itself (KU 64), i.e., its parts maintain the whole, while the whole also reciprocally maintains its parts. Nietzsche takes exception to Kant's claim, asserting that Kant has lost sight of the fact that this is "not a constitutive judgment, but only [a] reflective judgment."[39] There are purposes that are presupposed in how parts of an organism work together, but they are not purposive in the sense of having any reason authored by an intelligent designer. In this context, Nietzsche offers an account of the human understanding that suggests that the idea of "purpose" lies only in the human representation.

> When we speak about purposive concepts and causes, we thus mean: from a living and thinking being a form would be intended in which it wants life to appear. In other words, we approach through final causes not an explanation of life, but only its form. Now we grasp in a living being nothing but forms. The eternal becoming is life; through the nature of our intellects we grasp only form. In truth, no form can be given, because in each point an infinity dwells. Each thought unity depicts a line. A similar concept as the form is the concept of the individual. One names organisms thus as unities, as purpose centers. But they are only unities for our intellect. Each individual has an infinity of living individuals in himself. It is only a coarse intuition, perhaps first inferred from the bodies of humans. All forms can become thrown out, except for life! (BAW 3, 387)

Here we have Nietzsche's theory of the understanding as applied to the special problem of organic life, in which he states that our judgments actively structure and indeed create the identities we encounter. What we take as a given unity is really a multiplicity. There is a similarity here to Siddhartha Guatama, David Hume, and Daniel Dennett's teaching about the identity of the self, in the sense that a center of power is named, but there is no unchanging or permanent conceptual essence that is immanent within the organism. Outwardly, we name an organism by a concept, but the concept is not adequate to represent the perpetual changes of the parts and configurations that lie within. Furthermore, what we name as a singular organism is merely the surface of a power structure that belongs to a larger species. Since the distinction between individual and species is not absolute, Kant's claim that a "natural purpose" is unproblematically given oversimplifies biological inquiry. The problem of assuming the unconditional existence of the organism as a separate type of "thing" is a part of a larger problem of the origin of unity (*Einheit*), since it is precisely the unity of an organism that is in question, a unity that is used as the basis for the teleological proof for the existence of God.

Even beyond the problematic case of the identity of the organism, Nietzsche is clearly influenced by Lange's claims about the origin of unity and identity in general:

> We have seen how ancient materialism fell into absolute contradiction by regarding atoms as the only existent, though they cannot be the bearers of a higher unity, because without pressure and collision no contact takes place between them. But we also saw that precisely this opposition between manifoldness (Vielheit) and unity (Einheit) is peculiar to all human thought, and that it only

> becomes most obvious in atomism. The only salvation here, too, consists in regarding the opposition of manifoldness and unity as a consequence of our organization, in supposing that in the world of things-in-themselves it is resolved in some way unknown to us, or rather does not exist even there. In this way, we escape the inmost ground of the contradiction, which lies in the assumption of absolute unities, which are nowhere given to us. If we conceive all unity as relative, if we see that the unity is due only to the compilation (*Zusammenfassung*) from our thoughts, it follows that we have not grasped the inmost nature of things, but we have certainly made possible the scientific view. It fares ill indeed with the absolute unity of self-consciousness, but it is not a misfortune to get rid of a favorite idea for some thousand years.[40]

Lange's view that there is no absolute unity to be found in nature helps furnish Nietzsche with the methods he demonstrates in TSK. However, the problem of the organism offers a different variation of the problem of identity, since the parts of an organism work together. The phenomenal grasp of the organism offers no absolute unity or identity:

> The individual is an unjustifiable concept. What we see of life is form; we see it as individual. What lies within is unknowable (BAW 3 390).

In Nietzsche's assessment, a complex network of errors creates the idea of a unified purpose that was premeditated by a conscious being. This position that Nietzsche maintains is deeply indebted to Lange's view of the human understanding, as is Nietzsche's position that an intelligent creator within nature only exists according to human analogy.[41] Similarly, the singularity of the identity of an organism is a simplification based on the world of appearance (*Erscheinungswelt*), even if it empowers the subject by permitting a type of organization of the physical world. The fact that an organism exists indicates merely the capacity to survive. Nonetheless, there still is something inexplicable about the nature of organic life, since it proceeds without consciousness but still seems to exhibit a type of design without which it could not survive.

Nietzsche is concerned with eliminating teleological judgments in order to underscore that the processes of life do not ultimately point back to a "higher" knowing to which the movements of becoming could find correspondence. The innocence of becoming is thereby maintained. Nietzsche proclaims teleology to be only "an aesthetic product" (BAW 3 375), thus taking seriously the form-giving nature of reflective judgment. If Nietzsche is correct, teleological judgments are a way of simplifying and unifying the contents of the world, but this way of viewing things is not a type of knowledge.

Teleological judgments share the same terrain as aesthetic judgments in that they both indicate the workings of the creative imagination and they are both incapable of making determinative claims. Nonetheless, Kant winds up offering a teleological portrait to attempt to account for how organisms maintain and proliferate. Kant offers a reflective portrait of a "larger plan" of nature given by what he refers to as the special case of a "natural purpose," the organism. This is the point which Nietzsche attacks in TSK. Instead of invoking "natural purposes," which Kant claims are only valid for *our* power of judgment, Nietzsche offers an alternative explanation by pointing to mechanism and unconscious power. According to Nietzsche, there is no "big plan."

Kant asserts that when an organism can be viewed as both cause and effect of itself, i.e., its parts maintain the whole while the whole also maintains its parts in a reciprocal co-operation, it may be judged as a natural purpose, thus giving "objective reality to the concept of a purpose that is a purpose of nature rather than a practical (*praktischer*) one and hence giv[ing] natural science the basis for a teleology" (KU 65 375). Here Kant is using the word "*praktischer*" not in any ethical sense, but merely in order to put it in contra-distinction to an extrinsic (i.e., a "for-us-ness," not as it is in itself) teleological mode of thought. Nietzsche responds to Kant's supposition that we are entitled to claim that there are objective purposes by writing, "The concept of purposiveness is only the condition for existence. Nothing is said about the measure of reason disclosed within it" (BAW 3 384). Nietzsche thus stresses that the metaphorical nature of the teleological judgment has been lost sight of by Kant, so the "natural purposes" that Kant postulates are a product of anthropomorphizing, authored by a contingent human way of representing things. As Goethe had already stressed (BAW 3 376), the "things" themselves as given individuated unified organisms are only inserted by the human organization.

Nietzsche criticizes Kant for supposing that a premeditated unity and order may exist independently of the human imagination. Kant himself concedes that when we posit the existence of the natural purpose "nothing in the relationship of [our idea of] purposiveness gives it a similarity to the organism" (KU 65), fully recognizing that the process of assigning a possible cause to account for the apparent intelligent design is a reflective (i.e., non-cognitive) judgment. Kant thinks that such a way of presenting things, even if it cannot be knowledge, may be the best humans can do to make sense out of the existence of organic life. The problem of *how* the organism maintains and organizes itself without consciousness suggests an epistemological limit. Kant recognizes that our causal representations may be completely different than the mysterious unconscious processes of the organism.

Nietzsche does not lose sight of the problem of the limitation of our concepts, since our representations may be completely different from the natural organisms the concepts attempt to represent. This is why Nietzsche takes exception to Kant's subsequent move of assuming that the existence of organized life allows one to suppose from it that all of nature is governed by a unity of purpose:

> [T]his concept [of natural purpose] leads necessarily to the idea of collective nature as a system which operates according to the rule of purpose. (KU 67)

Nietzsche claims that this expanded view of teleological ordering principles of the universe could never be arrived at without a type of theological agenda lurking in the background. Nietzsche is careful in TSK to point out that Kant's view is only a hypothetical possibility, and one may just as easily revoke the presentation of teleological order by advancing the simpler alternative possibility of mechanistic explanation and casualism.

The hypothesis that "collective nature" has any purpose which could be deduced from the existence of an organism is bogus, according to Nietzsche. Likewise, the view that the organism is a "purpose of nature," an actual product of an intelligent conscious designer, is rejected in TSK. Such a claim is fiction, indicating no higher order:

> Laws of power, these and the capacities for life which act in the production and preservation would be the same thing: thus this is very unreasonable. (BAW 3 379)

According to Nietzsche the organization of experience via teleological judgments does not enable us to pass judgment on how things are independently of our schematization. Rather, the organism conceived as a "given" unity caused by an external conscious designer is itself only a reflective hypothesis, and is actually an extension of conceptual thought into a region that does not operate according to concepts:

> [T]he concept of the whole is our work. Here lies the source of representation of purposes. The concept of the whole does not lie in the thing, but in us. These unities which we call organisms are still multiplicities. There are in reality no individuals. Moreover, individuals and organisms are nothing but abstractions.[42] (BAW 3 379)

If the organic being could not "comprehend" itself as a whole it would not survive, as the conditions for existence are "built according to an infinite series of failing and half-successful attempts" (BAW 3 381). Yet Nietzsche suggests such "comprehension" between the parts and the whole of an organism can never be completely understood, since it is not a matter of the discursive understanding. Although Nietzsche rejects Schopenhauer's theory of the will, he still retains a residual view that Schopenhauer maintains at times: there is an unbridgeable gap between the lifeforce and representation. Kant had attempted to map out the limits of human reason in the natural sciences in the *Critique of Pure Reason*, but accounting for organic life poses its own special problem in terms of assessing the design in living organisms, a problem around which Schopenhauer constructs his entire philosophy. Therefore the *Critique of Judgment* is crucial to consider because there is a different kind of epistemological limit that Kant attempts to identify, the processes that power living organisms. Both thinkers attempt to present something that goes beyond the limits of what is knowable. Schopenhauer names it as the thing-in-itself. Kant names it as an artwork of God. Nietzsche's 1867-8 investigations suggest that neither thinker is correct, as both lapse into fictitious aesthetic presentations.

Because there are multiple forms of purposiveness, i.e., different forms of life whose relation between the whole and the parts hold nothing necessary other than the conditions for survival, the unity of the living organism which is presumed "given" must be ruled out in Nietzsche's view. Nietzsche suggests that we never can comprehend fully the inner animating force in any organism which appears organized without consciousness. The mere fact that organic life has sought out a form of purposiveness insofar as its cells work in conjunction with the whole only entitles us to claim that it has the conditions necessary for survival. It indicates nothing of a premeditated essential quality, nor can it be described as an "objective purpose," but originates as a non-cognitive reflective hypothesis about the nature of the organic. Since there are multiple forms of purposiveness, there is no essential "objective purpose" in regards to life. In response to considering an organic being as a product of design or objective purpose, Nietzsche writes:

> When a person pulls out a ticket from an urn and does not win the death lottery, this is neither purposive nor unpurposive, but by chance, i.e. it occurs without reflection preceding it. (BAW 3 376)

Thus Nietzsche claims that "what we call purposive is only what proves to be life-furthering" (BAW 3 380). In this regard, Nietzsche wants to suggest that the purposive can originate mechanically. One is tempted to jump to the conclusion of an intelligent teleological designer only because the individual does not see the vast number of forms which have perished due to a lack of conditions necessary for survival.

Kant's description of a purposiveness without consciousness suggests that the understanding cannot grasp the workings of organic beings: life can maintain itself without conscious deliberation, but according to this view there still is something mysterious about how the organism operates without consciousness. Yet Nietzsche draws very different conclusions from the mysterious nature of an organism, suggesting that it is not due to a design from an intelligent creator, but ascribes it to "chance [which] rules unconditionally, the opposite of purposiveness in nature" (BAW 3 386). Therefore according to Nietzsche's response, there is no indication of order bestowed by a designer, nor are we entitled to assume that there is an intelligent designer that directs order within the universe.

There still is something mysterious about organic life for Nietzsche, but there is a sense that there is no deep secret to the organism: it acts like a machine. There still is something about the living organism that exceeds conceptual grasp but this is due more to the surface play of human cognitive powers and their inability to grasp the perpetually changing nature of organic life. Human thought employs static concepts to represent the organism when the organism is never the same.

While Kant may still retain reflective hypotheses for teleological claims, he had generated a suspicion about any such claims which would disavow their regulative and reflective nature. Looming still larger in the culmination of his biological reflections is his recognition that we must adopt necessary fictions in order to investigate how life organizes and maintains itself.[43] Although Kant does not accentuate their fictioning nature to the degree of Nietzsche, he does realize that teleological judgments are only "valid" in the context of our power of judging. Kant's understanding of reflective judgment furnishes an important type of representation that allows us to represent how the world might be. It may even represent the noumena, but it does it in a way that need not be false nor lapse into speculative metaphysics, precisely because it is non-cognitive in nature. It would only become speculative in the Kantian sense of bad metaphysics if one were to assume that such a judgment is a claim to knowledge. Although there are different senses of aesthetic judgment in Aristotle and Kant, in this respect the use of teleological judgments by Kant has a similarity to Aristotle's view of the activity of the poet.[44] In the final analysis, reflective judgment has a peculiar status of a go-between, to envision a connection between the noumenal and phenomenal worlds. It is precisely this type of strange non-cognitive representation that marks the end of Kant's system and the beginning of Nietzsche's portrait of the human understanding.

In some respects, Nietzsche's TSK follows Schopenhauer's reflections on the inability of conscious thought to understand the blind, dark forces of life. The difficulty of the gap that results from the conscious representation of the unconscious persists as a theme for both thinkers, but Nietzsche ultimately parts with Schopenhauer, specifically rejecting Schopenhauer's claim that the will is the thing-in-itself that powers living organisms. Schopenhauer recognized the implications the *Critique of Judgment* had for delimiting the terrain of reason, and contributed to Nietzsche's investigation of this important dimension. Nietzsche never abandoned his interest in these vital concerns, for these questions condition his philosophy throughout his creative career: What is life? What is the meaning of life? What significance does discursive and conceptual thought have for the existence of life?[45] The ambiguous question of life itself indeed proves to be crucial for any philosopher who seeks to create a worldview that is in the service of life.

When seeking to understand how the unconscious co-operation between the parts and the whole of an organism are possible, Kant invokes his concept of natural purpose, but he still thinks that there is something that simply cannot be explained. Nietzsche thus quotes Kant:

> Through the concept of a mechanical lawfulness one can explain the universe, but no organisms would be explained. . . . It is impossible to represent the inhering substance. . . . Substance is only the outer appearance! (*außere Erscheinung*). (BAW 3 377)

Perhaps one might raise the objection that "life" is such a vague term that it can scarcely be pointed to as being capable of providing a unifying theme. Nonetheless there is no mistaking Nietzsche's deep concern with the inability of conceptual thought to render the workings of "life" translucent, in spite of any dialectical attempts at illumination. Moreover, "life" appears as that which both philosophy and history are to be in the service of in Nietzsche's thinking. This theme is further developed in Nietzsche's later critique of ethics, as he invokes a "hermeneutic of life" to determine whether the actions which ethics sanction are life-promoting or life-negating.

"Life" in Nietzsche's early dissertation is already described as "the eternal becoming" through which no light can shine by purposive causes, as "our intellects are too blunt to perceive the perpetual changing within" (BAW 3 387). Rather than positing another abstract metaphysical concept of ground, Nietzsche's accentuation of "the eternal becoming" announces a network of power relations and change. The unities which we graph onto the physical world are regulative fictions, fated to pass away like the moving river of Heraclitean flux. In this regard, it is not only organic life which exceeds our comprehension, but strictly speaking, any intuition of sense must import a form of the understanding through which the world is "seized" by mediation. The problem is that there are neither organic nor inorganic unities which are unproblematically "given."

Nietzsche also indicates in TSK that "Purity would only be known in the mathematical" (BAW 3 382). The problem with the mathematical in this view is that it can only amount to tautologies that human beings invent.[46] In this view, humans create formal "truths" via representation but, strictly speaking, they cannot refer to anything outside of consciousness without the assumption of a "given unity" in the physical world.[47] If this view were correct, furnishing "the given" would always prove to be an artistic insertion by the human organization. Once again, connecting the *a priori* precision of logic and mathematics to the world is the problematic gap that manifests itself. Humans must make the inexact world *fit into* our ways of representing it.

The inability to apply the certainty of mathematics to the workings of life is not insignificant in that it indicates that we also require a creative act, one that is able to bridge the cleft between the ideal and the real. By the use of language we refer to the identities of organisms and things, but these are only names that are authored for the sake of convenience of speech, since organisms and things may change perpetually, yet retain the same name. Thomas Reid recognized that this convenience of speech seemed to undermine the idea of an absolute identity, since it only yields a sense of identity in a loose sense.[48] Language could not work as an organizing principle if we had to create a new name whenever a phenomenal object undergoes a change, especially since the world seems to be in a state of perpetual flux: the world is *always different*. Thus human beings import identities into the world in a very basic sense, one which empowers them by permitting an organization of the world as it appears to them, but such an activity is still grounded in this type of "all too human" creative perspective. Nietzsche's view that human beings project stability and permanence into the world coheres with his Heraclitean conception of tragic thinking, a view that suggests an unavoidable creative perspective.

For Nietzsche there is a sense that tragic thinking is closest to representing the nature of the world because it dispenses with the nostalgia for a moral world order. Such a tendency also relinquishes the concept of a "higher knowing" to which "things" in the world could in principle be referred back to for sanction by virtue of correspondence.[49] Kant's recognition that we are incapable of providing determinate judgments on the "larger plan" announces a humbling of the human powers of cognition. Yet relegating our claims as to how the "larger plan" works to a judgment which resembles a type of aesthetic judgment opens other possibilities for explanation. If teleological order is just a way of viewing things, does such a view help or hinder the investigation of nature?

Thus on the one hand we have a schematization of experience where any claims to knowledge cannot make reference to "the way things are" independently of how they appear to us. On the other hand, we are capable of proceeding with certainty only insofar as we can grasp "truth" in the purely formal and mathematical. Nietzsche suggests that any concept that purports to represent the world that exists outside the subject arrests the flow of becoming, altering the world in the very representation of it. If this view applies for all phenomena, it would seem that the transcendental subject can never achieve truth as correspondence without invoking a claim to which it is never entitled, i.e., that the "external" world is as we think it—appearing as a finite representation in time and space. Much later, Nietzsche seems to reiterate the same position in the *Will to Power*:

> Mathematics is possible under conditions under which metaphysics is never possible. All human knowledge is either experience or mathematics. (WP 530)

This passage appears in the context of a larger criticism which is specifically directed toward Kant, and it accentuates the gap between the real and the ideal.

Nietzsche quotes the following passage from section 68 of the *Critique of Judgment* twice in "*Teleologie Seit Kant*":

> We have complete insight only into what we can make and accomplish according to concepts. (BAW 3 381, 383)

Nietzsche seems to advance the view that we are able to know "things" conceptually only insofar as we refer to our own networks of regulative fictions, i.e., we are only able to cognize that which we ourselves make. If this view is correct, making the jump from the human understanding to cognition of the world itself requires that we reconceive the history of knowledge as anthropomorphic perspectivism. Because we are form-givers at bottom, humans are artists in a general sense much more than we are consciously aware.

Even as early as 1868, Nietzsche's critique of the power of the human understanding in TSK underscores the process of making the strange "not-strange," making the infinite finite. By examining these early notes, the depth of Nietzsche's thinking concerning power, anthropomorphic fictioning, becoming, and life is revealed. In the larger picture, there is the danger of overestimating the importance of this early thesis that was never completed. Yet the theme of the necessity of making the infinite finite is a problem which he raises still again toward the very end of TSK. Quoting Hamann's letter to Lindner on Kant, he writes:

> His errors are blind children, who are thrown out by a hasty bitch. . . . He promotes himself to judge on the whole world from the parts with which he is familiar. But for that reason it is a knowledge which is nothing more than bunglery. To unlock the whole from the fragments is thus equivalent to unlocking the unknown from the known (BAW 3 392).

"To unlock the whole from the fragments" would never appear to be justified, as the fragments themselves have been taken as "given" evidence of an intelligent designer. However, the fragments themselves are not wholes, i.e., the fragments are by no means unconditional, but are inserted by the human organization. In other words, Nietzsche is stressing that we see organisms as unities, but these unities to which we ascribe design are really multiplicities. We use this fictitious unity to project an idea of an intelligent cause into nature, and then we make the further conceptual leap that the universe itself exhibits a unified teleological order.

By concluding the entire critical enterprise with a critique of the power of primordial parting, Kant initiates a non-cognitive, non-ethical type of thinking (*Denkungsart*). Nietzsche's assaults on epistemology and ethics may be interpreted as an aesthetic response to this "other" kind of thinking. Invited by Nietzsche's early metaphysical proximity to the *Critique of Judgment*, the recognition that art is necessary to conclude Kant's critical enterprise suggests a deeper affinity between the two thinkers, one that recasts the artwork not in terms of a philosophically isolated region, but rearranges the philosophic terrain in terms of a piece of art.

2. Introduction to the Translation of "Teleology Since Kant" (1868)

While recovering from his horse accident during military service in early 1868 Nietzsche describes a "half philosophic, half natural science"[50] doctoral dissertation to Rohde, which was to be entitled "On the Concept of the Organic Since Kant."[51] Although Nietzsche indicated in a draft of a letter to Anton Klette in February 1869 that he had abandoned this project, the notes which constitute the dissertation draft are the longest sustained inquiry we have on Kant from any of Nietzsche's writings. It is very likely that Otto Kohl's dissertation on Kant may have helped suggest this theme to Nietzsche in April 1868.[52] There does not seem to be any evidence that Nietzsche had the dissertation plan prior to this time, even though it converged and overlapped with his Democritus reflections.[53]

Appearing as "*Teleologie Seit Kant*" under "Juvenilia" in the Musarion edition (*Gesammelte Werke*, 1920-9), the early dissertation draft was republished in a slightly different form in the Beck edition (*Beck'sche Ausgabe Werke*, 1933-40). The editors of the latter edition had planned to publish everything which Nietzsche wrote, but this project was abandoned in World War II after only five volumes appeared. At the close of the twentieth century the Beck edition finally became widely available, as it was republished as *Friedrich Nietzsche: Frühe Schriften* in 1994. In 1999 the material that makes up *Teleology Since Kant* finally appeared in *Nietzsche Werke,* a supplementary volume of the *Kritische Gesamtausgabe*.

In its broadest outlines, "*Teleologie Seit Kant*" responds to Kant's postulation of the "natural purpose" in the "Critique of Teleological Judgment." In *that* part of the *Critique of Judgment,* Kant offers the culmination of his biological reflections directed at the unconscious co-operation between the parts and the whole of an organism. Kant clearly suggests that the "dark side" of the organism is inaccessible to the discursive intellect and consequently he thinks that the organization of nature "has nothing analogous to any causality that we know."[54] Thus Kant stresses that we think of "natural purposes" only by analogy, so the mode of representation of organisms is transferred from things that we actually do know by an image creating faculty which, strictly speaking, cannot yield cognition. Yet Kant loses sight of the analogical nature of the representation of organic beings and moves beyond the organism to offer an expanded representation of teleological order, to consider all of nature as purposively arranged.

In response to Kant's recognition that the discursive intellect cannot render the workings of "life" translucent, Nietzsche offers various attempts by "philosophers of nature" (i.e., Schopenhauer, Goethe) to deal with Kant's insight. Ultimately, the project investigates what Nietzsche would later refer to in his 1869-70 essay "On the Origin of Language" as that "remarkable antinomy" from the *Critique of Judgment* which makes "the correct understanding of language possible"—that "something can be purposive without consciousness."[55] Clearly Nietzsche agrees with Kant's view that there is something inaccessible to the discursive intellect which somehow allows a living being to proliferate and maintain itself. Yet Nietzsche parts from Kant by claiming that there is no compulsion to assume that any concept or rational design had to precede the existence of an organism to make it possible. Rather, Nietzsche claims that organisms are nothing but a function of power relationships: we only see organisms that have had the capacity to survive, but this indicates nothing about any external designer.

It is worth noticing that Kant seeks recourse to teleological judgment only after attempts to explain life mechanistically have been (in his view) exhausted. Kant offers a peculiar "end" of the critical project which offers no doctrine,[56] but only a critique of the power of reflective judgment: a non-cognitive, non-ethical way of thinking. It is this juncture of the Kantian project which explicates the imagination's capacity for free invention. Nietzsche responds to this movement by suggesting that human beings are "deeper" artists than they are aware, since they are the authors of the ideas of "force, matter, individuals, law, organism, atom, and purposive causes. These are not constitutive judgments, but only reflective judgments."[57] Elements from Nietzsche's later "falsification thesis" clearly are already present in this work, as he suggests that the human understanding invents identities that do not exist "outside" of human consciousness.

In the translation which follows, I generally follow the Beck version, indicating where it has slight variations from the Musarion edition. The former lists Nietzsche's citations of pagination from Kant, Schopenhauer, and Goethe, while the latter omits these details. This may not seem terribly significant, but the correspondence of Nietzsche's page citations of the *Critique of Judgment* with the Rosenkranz edition of 1838 indicates that Nietzsche did read Kant firsthand, in spite of also relying heavily on Kuno Fischer's lectures on Kant (among other sources) in the *Geschichte der Neuern Philosophie*.

There is also a similarity of Nietzsche's dissertation draft to Hume's *Enquiry Concerning the Human Understanding*, since Hume utilizes Democritus's atomistic heir, Epicurus, as a mouthpiece through which he undermines claims about any order bestowed upon nature by an intelligent designer. Using Epicurus as a literary device, Hume writes: "You forget that this superlative intelligence and benevolence are entirely imaginary, or at least without any foundation in reason; and that you have no ground to ascribe to him qualities, but what you see he has actually exerted and displayed in his productions." Similarly, Nietzsche's assault on teleology is partially indebted to the ancient schools of atomism.[58]

The notes which follow demonstrate a unity of thought, but still do not constitute a polished work. I have consulted the only prior English translations of *Teleology Since Kant* of which I am aware. One of these appears in an appendix of Claudia Crawford's *The Beginnings of Nietzsche's Theory of Language* (de Gruyter Press, 1988).[59] By translating *Zweckmässigkeit* as "purposiveness" rather than "expediency," the literal stress on "purpose" which is present in the German is reflected in my translation. Elsewhere I have argued that there are other problems with the prior translation of this work, particularly in regard to the issue of "casualism," a view which Nietzsche advances to explain organic life.[60]

I have inserted brackets <[ ]> within the text to identify the sections where Nietzsche quotes from Kant's *Critique of Judgment*, along with the corresponding section numbers; where Nietzsche refers to section numbers himself I have copied parentheses <( )> from the text. This work provides an additional part for a genealogy of Nietzsche's philosophy. A sustained critical rigor is prominent in this early phase which offers us Nietzsche's earliest statement of a theory of the understanding, a theory which begins by critiquing Kant's portrait of reflective judgment.

3. Teleology Since Kant (1868)

Kant wants to show that "a necessity exists for us to think of organic bodies as premeditated—to think of them as coming into being according to concepts of purpose."[61] I can grant that this is only one way to attempt to account for natural bodies, to explain them to oneself according to teleology.

Human experience proceeds by analogy to demarcate concepts of purpose which still are accidental: the origin of purposiveness in natural bodies is not a premeditated product. In other words, the origin of organic bodies is due to the fortuitous working together of talent and fate, the draw of the lottery. From an infinite number of chances to generate organic bodies, there must also be the emergence of some bodies that turn out to be advantageous or purposive.

The necessity of which Kant speaks no longer exists in our time: but one is reminded, however, "that even Voltaire held the teleological proof to be irrefutable."[62]

Optimism and teleology go hand and hand: both set out to deny that the unpurposive is something which is actually unpurposive. To take arms against teleology in general is the task at hand—to offer a proof of unpurposiveness. Through this it would only be shown that the highest reason had worked sporadically, and that there is also a terrain of lesser reasons given. There is thus no unified teleological world, but rather, a creative intelligence.

The assumption of a unified teleological world would have been made only according to a human analogy: why can the purposive not be an unconscious creative power, i.e., which nature produces? One thinks of the instinct of the animal. This is the standpoint of natural philosophy.

One no longer posits the knowing as lying outside of the world.[63]

But we remain stuck in metaphysics and must bring in the thing-in-itself.

Finally a solution can be reached from a strictly human standpoint: the Empedoclean, where the purposive appears only as an accident arising from much unpurposiveness.[64]

Two metaphysical solutions are attempted: one is a coarse anthropological attempt that idealistically posits a human conception which is external to the world; the other, which is also metaphysical, takes flight into an intelligible world, in which the purpose of things is immanent. The purposive is the exception. The purposive happens by chance.

The purposive reveals itself as being completely without reason.

One must eliminate every theological interest from the question.

Teleology Since Kant.

Nature philosophically considered.

The organism as a simple idea is shattered in the multiplicity of its parts and its condition, but it still continues as unity in the necessary unification of its parts and functions. This unity of the organism is created by the intellect.

"The purposiveness of the organic, the lawfulness of the inorganic are brought into nature by our understanding."[65]

This idea, expanded, yields the explanation of an externally conceived purposiveness. The thing-in-itself must "indicate its unity in the agreement of all its appearances with one another."[66] "All the parts of nature accommodate each other because a single will is there."[67]

But in opposition to the whole theory forms that horrible struggle of the individuals (which also manifest themselves as an idea) and the species. The explanation thus presupposes a teleology throughout: it does not exist. The difficulty is precisely the connection of the teleological to the nonteleological world.

The status of the problems.
Kant's rejection of attempts at solution.
Solutions of the natural philosophers.
Critique of Kant's view.

The question has a similarity to that of seeking after the freedom of the human will, in that it sought its solutions in the realm of the intelligible world, while it overlooked another coordinated possibility.

There is no question that exists which requires the necessary assumption of an intelligible world for its solution.

Teleology: intrinsic purposiveness. We see a complicated machine, which maintains itself and can not contrive any other structure that could construct it more simply. But that only means that the machine maintains itself, thus it is viewed as purposive. A judgment about the "highest purposiveness" does not belong to us. We could at best infer that there is a reason, but we have no justification to indicate whether such judgments about purposiveness exhibited within the machine are higher or lower.

An externally conceived purposiveness is an illusion. In opposition to it, we recognize the method of nature as it generates such "purposive" bodies, through a senseless method. Consequently, purposiveness proves itself to be only a capacity for life, that is to say, as the *cond. sine qua non* for life. Chance can hit upon the most beautiful melody.

Secondly, we know the method of nature, as it would maintain itself as a purposive body: it occurs with senseless thoughtlessness.

But teleology throws out many questions which are unsolvable or which have not been solvable until now.

The organism of the world, as well as the origin of evils, do not belong here, but are examples that originate from the intellect.

Is it necessary to put teleology in opposition to the explained world?

It only indicates another reality of a domain that has fixed limits. The position that opposes the rejection of teleology claims that teleology allows the logical laws themselves to be disclosed further on higher levels; but we are not permitted to speak of logical laws.

Purposiveness.

We see a method for the attainment of purposes, or more correctly: we see *existence* and its means and infer that its means are purposive. Therein lies still not the recognition of a high, let alone the highest degree of reason.

We are astonished of the complexity of organic beings and then presume (according to a human analogy) a special wisdom lying within. The wonder in us is actually organic life: All the means it uses to maintain itself we call purposive. Why do we suspend the concept of purposiveness in the inorganic world? It is because we have pure unities here, but we do not have unities whose parts belong with each other working together.

The elimination of teleology has a practical value. Only then is it possible for the concept of a higher reason to be removed: thus we are already satisfied.

The evaluation of teleology in its value for the ideas of the human world:

Teleology is, like optimism, only an aesthetic product.

The strict necessity of cause and effect rules out the purposes in unconscious nature. Because purposive representations are not generated in nature, they must lie here extrinsically, as outside the causality and consequently they must become viewed as a motive inserted here and there, wherethrough even the strict enduring necessity would be broken. Existence is perforated with miracles.

Teleology as purposiveness and effect always pushes conscious intelligence further. One asks for the purpose of this isolated interference and one is confronted with the purely arbitrary.

"Order and disorder are not in Nature."[68]

"We ascribe effects to the accidental, whose connection to the causes we do not see."[69]

Things exist, consequentially they must *be able* to exist, that is to say, they must have the conditions for existence:

When the human being constructs something, i.e., wants to make something capable of existence, he thus reflects under what conditions this can happen. After the fact, he names the conditions for the existence of his fabricated work *purposive*. Therefore he also calls the conditions for the existence of the thing *purposive*—that is to say, only by virtue of this assumption would they originate as a human creation.

When a person pulls out a ticket from an urn and does not win the death lottery, this is neither unpurposive nor purposive, but as they say, *by chance*, i.e., without any thought preceding it. But this is the condition of his continuing existence.

Is it true that Democritus would hold that the origin of language arose out of convenience?[70]

"The organization of nature has nothing analogous to any actual causality that we know." Kant says this about the organism in the Critique of Teleological Judgment (p. 258).[71] [KU 65].

"An organized product of nature is that in which all is a purpose and reciprocally a means" (p. 260, [KU 66]).

"Each living thing", says Goethe, "is not a singular thing, but a plurality: it is itself insofar as it appears to us as an individual, though it still continues as a collection of living independent beings." (Goethe, B. 36, p. 7, Introduction of *Structure and Change of Organic Nature*).[72]

It is very important for Goethe (for natural philosophy in general, intuition of the power of judgment)[73] (B. 40 p. 425) for the origin of his natural philosophy out of a Kantian principle.

"What the understanding knows through its concept of nature is nothing but the effect of moving force, i.e. mechanism."[74] "That which cannot be known through a merely mechanistic understanding does not belong to exact natural science."

"Mechanistic explanation means explanation by external causes."

"Specification from external causes is not able to explain nature." "But we have nothing without cause." Thus we must think inner causes, namely, purposes which still are representations.

"A way of viewing things is still not knowledge."

"The principle of such a necessary way of viewing things must be a concept of reason." "The only principle of this kind is natural purposiveness."

"Through the concept of a mechanical lawfulness one can explain the architecture of the world, but no organisms would be explained."

"It is impossible to represent natural purposiveness as inhering in the matter."

"Matter is only outer appearance" [KRV a359].

"The purposiveness of things can always only be valid in relation to an intelligence, with which the insight of a thing is in correspondence." And certainly "either in our *own* or in a *foreign* intelligence is the ground on which the thing itself lies. In the last case, the intention of the existence of the things would be revealed in the appearance." In other cases, only our representation of things would be judged purely as purposive. This last kind of purposiveness relates itself only to the form ("in the simple observation of the object the imagination and intelligence are harmonized").

"Only through the mechanistic origination is the thing cognizable."

A class of things is not knowable.

We understand only mechanism.

Through mechanistic origination a thing is cognizable, but we do not know whether this representation gives us something completely different. It is determined by our organization that we are only able to understand a mechanistic generation of the thing.

Now there is also (says Kant) a compulsion in our organization which makes us believe in organisms.[75]

From the standpoint of human nature:

We know only mechanism.

We do not know the organism.

But mechanism, as is the case with the organism, does not give us the thing-in-itself.

The organism is a form. We overlook the form's multiplicity.[76]

The organic body is a substance in which the parts are bound together purposively. Therefore, we long for causes which bind together the parts of a substance purposively; i.e., Kant says (sec. 65, 66) "organizing causes, which must actually become thought of as working according to purposes."

But therein lies a fault. It is only necessary to point out another possibility to remove Kant's presentation concerning the *compulsion* to think that living things are consciously organized. Mechanism bound with casualism yields this possibility.[77]

That which Kant claims—he claims according to a bad analogy: since it namely follows his acknowledgement that nothing within the organism has a similarity to the purposive relationship which is represented. (KU 65)

The purposiveness originates as a special case of possibilities: an innumerable quantity of forms originate from mechanistic combination: from these innumerable forms some can also arise which have the capacity for life.

The hypothesis is living beings are able to originate out of mechanism. Kant *denies* that.

In truth, it stands firmly that we only cognize the mechanistic. That which is beyond our concepts is wholly unknowable. The origin of the organic is thus hypothetical, as we represent it to ourselves as if a human understanding would be present. Yet the concept of the organic is only human: One must point out that the organic only exists as a human analogy. The ability to live originates under an enormous number of possibilities incapable of life. With this insight, we bring ourselves closer to resolving the problem of the organism.

We see that many capacities for life originate and would maintain themselves and then observe the method.

Laws of force, these and capacities for life which act in the production and preservation of the organism would be the selfsame: thus this is very unreasonable.

But this is the assumption of teleology:

"The idea of *effect* is [. . .] the *concept* of *the whole*" (KU 65). "In organisms the active principle is the [. . .] idea of created effect" (KU 65).

However, the concept of the whole is our work. This is where the source of the representation of purposes lies. The concept of the whole does not lie in the thing, but in us. But once again, these unities which we call organisms are still multiplicities. There are in reality no individuals. Moreover, individuals and organisms are nothing but abstractions. They are unities manufactured by us into which we transfer the idea of purpose.

We assume that there would be a unitary force which brings about organisms. Then the method is to observe this force of the organisms which allows them to create and preserve themselves.

Here it is demonstrated that what we call purposive is only that which proves itself to be capable of living.

The secret is only "life."

Is this also only an idea conditioned by our organization?

"The raging wastefulness astonishes us." Schopenhauer says (*World as Will and*

Representation volume II, p. 375 [Chapter 26]): "The works of nature cost her no trouble." Therefore she is indifferent to their destruction.

Schopenhauer thinks that there is an analogy to the organ (WWR a.a.O.) to "The movement of the will, the motive that directs *it* (final cause)."

Goethe's Attempt

The *Metamorphosis* [*of Plants*] has to do with the explanation of the organic according to the *efficient* cause.

Each *efficient* cause ultimately borders on something which is unfathomable. (That even proves that this is the correct human path).

Therefore one does not ask for final causes from inorganic nature, because there are no individuals there, but mere forces to be observed. In other words, because we can resolve everything mechanistically, we no longer believe in purposes.

"[we have] complete insight only into what we can ourselves make and accomplish according to concepts" [KU 69]

A false opposition

When in nature only mechanistic powers rule, the purposive appearances are only fictitious—their "purposiveness" is only *our idea.* Blind powers act unintentionally—thus they cannot produce purposiveness.

The capacity for life is constructed from an unending chain of failing and half-successful attempts.

The life of the organism proves no higher intelligence: In general, it exhibits no continuous measure of intelligence.

The existence of the organism points only to blindly acting forces.

1. Elimination of the expanded representation of teleology.
2. Limits of the concept. The purposive in nature.
3. Purposiveness is equivalent to the capacity for existence.
4. Organisms as multiplicities and unities.

"The *representation* of the *whole* as cause is thought according to purpose" [KU 77] . . . . but the "whole" itself is only a representation.

Kant [KU 77]:

Is it "possible that organisms originate purely mechanistically? [Kant says] It is impossible for us to infer that life is mechanistically generated." Why? The understanding is discursive, not intuitive: "it can grasp the whole only in its parts and put them together." But in organisms "the parts are conditioned through the whole. Now the understanding fails to proceed to the whole, for the whole is not given in the intuition but rather it is only given in the representation. The representation of the whole shall thus condition the parts: the representation of the whole as cause, in other words, as purpose."

"When grasping the whole out of the parts, the understanding proceeds mechanistically. When grasping the given parts from the whole; it can only derive them from the *concept* of the whole."

In short, an intuition of the organism is lacking.

A polemic in accordance with nature.

First of all it would be denied that the whole of an organism really exists. In other words, the concept of unity would prove to be inserted by the human organization. Thus from there we are not allowed to go beyond.

In organisms, not only are the parts conditioned by the whole, but also the whole by the parts. Thus if others are mechanistically generated from the organism, they must also be mechanistically derivable.

It is given that we can only see one side. Now the parts are next observed and broken down into their parts. Thus one arrives at the cell.

Under the view that organisms are generated mechanistically we can suppose that if a purposive concept would also be active, it would happen in spite of its creation through mechanism (as Kant acknowledges). Thus there must be a mechanism that can be traced.

The generatio aequivoca unproved (*unerwiesen*).[78]

Purposive causes, just as mechanism, are human modes of intuition. Purity would only be known in the mathematical.

The law (in inorganic nature) is, as law, something which is analogous to purposive causes.

"That in nature which is not composed by pure mechanism . . . is not an object of the understanding."

It allows only the strictly mathematical in nature to be explained.

"To explain things mechanistically means to explain things according to external causes." (This definition would be introduced, then later set into opposition against the internal).

To explain things mechanistically means much more. . . .

"We have complete insight only into what we can ourselves make and accomplish according to concepts" [KU 68].

Thus one can have complete insight only into the mathematical (the formal view). In all other cases, one stands in the unknown. To overcome this, human beings invent concepts, but these are only the sum of apparent qualities combined together, which do not get a hold of the thing.

These invented concepts include force, matter, individuals, law, organism, atom, purposive cause. These are not constitutive, but only reflective judgments.

Under mechanism, Kant understands the world without purposive causes: the world of causality

We cannot represent crystallization to ourselves without the idea of effect.

The origin and maintenance of organic beings—to what extent do they have to do with purposive causes?

Compare Kant on purposes of nature regarding the creation and preservation of individuals and the species with ([KU] 62). Then Kant substitutes the concept of a *thing* at ([KU] 63) and loses sight of the general forms of purposiveness.

The chance happening of a thing's form in relation to reason (is also found with the crystal).

"A thing exists as natural purpose when it is from itself both cause and effect" [KU 64]. This sentence is not deduced. A single case is given.

The derivation that organisms are the *single* kind of natural purpose is not successful.

On the contrary, in nature a machine also would already be led by purposive causes.

The concept of purposiveness is only the capacity for existence. Nothing can be said about the measure of reason disclosed within it.

"It is something different, says Kant [KU 67], to consider a thing according to its intrinsic form and to regard the existence of this thing as a purpose of nature. Therefore, the unpurposive method of preservation and reproduction of an organism does not continuously struggle with the purposiveness itself."

In opposition to this, to say that this organism is purposive is the same as saying this organism is capable of life. Thus it is not the case that the existence of this thing is a purpose of nature, but what we call purposive is nothing other than that which we find capable of life in a thing and as a result, we call those conditions purposive.

Whoever describes nature's method toward preservation as being unpurposive, observes even the existence of a thing as a purpose of nature.

The concept of natural purpose remains fixed only in the organism. "But" says Kant: "this concept leads necessarily to the idea of collective nature as a system governed according to the rule of purposes" [KU 67]. "Through the example which nature gives to us in her organic production one is justified to anticipate nothing from her and her laws, except that which is purposive for the whole" [KU 67].

This reflection comes only from the position in which one

1. disregards the subjectivity of purposive concepts
2. conceives nature as a unity,
3. conceives nature also as capable of a unity of means.

p. 267[79]

"If one thus in the context of natural science brings in the concept of God in order to make purposiveness in nature explainable, and then afterwards needs this purposiveness again to prove there is a god, then there is in neither science nor theology an inner stability and a deceptive circle brings each into uncertainty, through which their borders are allowed to jump over each other" [KU 68].[80]

To exclude the method of nature from preservation and generation of organisms in general is not the Empedoclean insight, but probably the Epicurean. But that view presupposes that chance can throw together organized being: while here the disputed point lies. From letters a tragedy itself can be thrown together (against Cicero), out of meteorites an earth: but the question is what "life" is, whether it is a merely an ordering and form principle (as in the case of tragedy) or something wholly different.[81] Against this it is given intrinsically that for organic nature in the conditions of the organism, no other principle exists other than those of inorganic nature. The method of nature in her treatment of things is the same. She is an impartial mother, equally harsh to her inorganic and organic children.

Chance rules unconditionally, the opposite of purposiveness in nature. The storm which pushes things around is by chance. *That* we can know. Here arises the question: Is the force which makes the things the same one as that which maintains them? etc.

In an organic being the parts are purposive for its existence; in other words, it would not live if the parts were unpurposive. But with that not a single part is yet determined. The part is a form of purposiveness, but it is still not settled whether this is the single possible form. The whole conditions nothing necessary within the parts, while the parts necessarily condition the whole. Whoever maintains the former view also maintains that there is the highest purposiveness, i.e., from among the different possible forms of purposiveness of the parts, the highest purposiveness is selected: in so doing, one accepts that there is a hierarchy of purposiveness.

Now which is the idea of effect? Is it life under its own necessary conditions? Is it an idea of an effect shared in common by all organisms?

So is it life in one form under the necessary conditions for that purpose? But the form and the conditions here happen together. In other words, if the form as cause would be posited, the measure of purposiveness would thus be equivalent to the cause thought within—because life in a form is quite certainly an organism. What is the organism other than form, formed life?

But if we say the parts of an organism would not be necessary, we would also say the form of the organism is not necessary. In other words, we posit the organic where it is nothing outside of the form. But outside that it is still simply life. Thus our principle is: for life there are different forms, in other words, multiple purposivenesses (*Zweckmäßigkeiten*). Life is possible under an astonishing number of forms. Each of these forms is purposive: but because a countless number of forms exist, there is thus also a countless number of purposive forms.

In human life, we create hierarchies of purposiveness: we are the ones who first posit occurrences as being equally "reasonable," [projecting the concept of reason] only when an entirely narrow selection occurs. If the human being in a complex situation finds a single purposive path, thus we say he acts reasonably. But if he wants to journey in the world and takes whatever path he pleases, he thus acts purposively, but still not according to reason. Therefore a reason for being does not disclose itself in the "purposiveness" of organisms.

That "which is thus the idea of the effect of cause" is only the form of life. Life itself can not be thought of as purpose, because life is already presupposed in order to be able to act in accordance with purposes.

When we speak about purposive concepts and purposive causes, we thus mean: from a living and thinking being a form would be intended from which life would appear. In other words, we approach through final causes not an explanation of *life*, but only its *form*.

Now we generally grasp in a living being nothing but *forms*. The eternal becoming is life; through the nature of our intellects we grasp only forms: our intellect is too blunt to perceive the perpetual change: that which is knowable to it, is called form. In truth, no form can be given, because in each point an infinity dwells. Each unity (point) thought depicts a line.

A similar concept as the form is the concept of the individual. Organisms are thus spoken of as unities, as purpose centers. But they exist only as unities for our intellect. Each individual has an infinity of living individuals in himself. The unity is only a coarse intuition, perhaps first inferred from the bodies of humans.

All "forms" can become thrown out but life!

By "the idea of the whole as cause" it is said that the whole conditions the parts, nothing further: since it is self-evident that the parts make the whole. When one speaks of purposive causes, one only means that the form of the whole floats before it in images of the parts—that a form could not have originated mechanically.

Life, including generation, is not to be confined under purposive causes. Life as a "self-organizing" being [KU 65] is derived arbitrarily by Kant.

Does one need purposive causes in order to explain that something lives? No, only to explain how it lives.

Do we need purposive causes to explain the life of a thing?

No, to us "life" is something wholly dark, through which no light can shine by purposive causes. We seek only to make the forms of life clear to ourselves. When we say "the dog lives" and then ask "Why does the dog live?" that cannot belong here. For here we have taken "life" as equivalent to "being there." The question "why something is" belongs to the externally conceived teleology and lies completely out of our area (childish anthropomorphic example[s] also by Kant).[82]

We cannot mechanically explain the power that makes a dog a living being.

The form is all that appears on the visible surface of "life." The contemplation according to purposive causes is thus a contemplation according to forms.

Actually, we are also required to question the springing up of crystal according to purposive causes.

In other words: teleological observation and how we view organisms do not synchronize with each other, but rather teleological observation and how we view organisms happen according to forms. Purposes and forms are identical in nature.

When the natural philosopher holds that an organism can be generated out of "accident," i.e., not according to purposive causes, he thus means that the form is added after. He asks himself only what "life" is.[83]

What justification do we have to consider the mode of appearance as a thing, for example, to conceive of a dog as pre-existing? The form is something only for us. We think of it as cause, so we lend the appearance the value of the thing-in-itself.

"Purposiveness" is only said in the context of "life." Thus it is not in reference to the forms of life. Thus in the concept of purposiveness there does not lie any recognition of rationality.

What shall be conceived of as the "idea of the effect of a cause" cannot be life, but only the form, i.e., a mode of the appearance of a thing would be designated as pre-existing and real.

A thing lives—thus its parts are purposive: the life of the thing is the purpose of its parts. But there are countless different ways to live, i.e. countless forms, that is to say, parts.

The purposiveness is not absolute, but is very relative: from other perspectives, it is often seen as unpurposiveness.

Purposive cause means: the idea of a whole would be indicated as cause—a phenomenal form (*Erscheinungsform*) would be indicated as real and pre-existing. The concept of the whole relates itself only to the form, not to "life."

1. It is not the case that "a 'life' is created, so a form for life must have been chosen,"
2. but rather, "under a succession of forms shall a 'life' appear."

It is impossible for the concept to grasp life: thus it does not belong to "the idea of the whole."

On the possibility of an origination of organisms by "chance," by "unpurposiveness" (mechanism). Kant considers the possibility, but denies the possibility of knowing it.

The method of nature treats the organic and inorganic equally. When the possibility of mechanism is there, thus there is also the possibility of knowledge. But "our understanding is discursive." But that suffices also when mechanism is explained.

The concept of an individual is inadequate. What we see of life is form; we see life forms as individuals. What lies within them is unknowable.

Procreation is not confined under purposive causes, since purposive causes ask "to which purpose shall a being become?" This belongs only in an external teleology, i.e., in a system of natural purposes.

A system of natural purposes has the following principles against itself:

1. The subjectivity of purposive concepts is taken as objective.
2. Nature is conceived as a unity.
3. And nature is conceived as capable of a unity of means.

Is a thing therefore not purposive because it is mechanically generated? Kant holds this. Why can not chance bring forth purposiveness? He had it right when he said that the purposiveness lies only in our idea.

Life occurs with sensation: thus as a condition for the "organic" we observe sensation.

Life is conceived as conscious, i.e., to exist according to human analogy.

The question concerning organisms is the following: where does the view that conceives nature as similar to the human view originate? From a deficiency of self-consciousness?

We can not represent "life" to ourselves, i.e., the sensory, growing existence, in any other way than one that is analogous to the human. Human beings recognize something in nature similar to the human analogy and some which are unfamiliar for which an explanation is asked.

I have observed that one also often thinks continuously in sleep: a chance awakening instructs one about scraps of thoughts which still linger in one's head.

Do we understand the unconscious co-operation of individual parts with the whole of an organism?

In inorganic nature, i.e., in the architecture of the universe, lawfulness and purposiveness are very much indeed to be thought of as consequence of mechanism. "In that Kant saw a *necessity of a plan*, the opposite of chance" (Kuno Fischer, *History of Modern Philosophy*, Volume 3, chapter 7, section 2).

Highly noteworthy point "to me it seems one could say in a certain state without audacity—give me matter and I will show you how a world from that shall be generated—etc." (Kant, *Universal Natural History and Theory of The Heavens*, Second Part, Preface).[84]

What Hamann said about Kant's optimism (*Attempt at Some Observations About Optimism*) is valid for optimism in general: "his notions are blind puppies, who are thrown out by a hasty bitch—He calls upon the whole itself in order to judge the parts of the world. For that purpose it is a knowledge that is no longer even bunglery. To infer from the whole to the fragments is the same as going from the unknown to the known" (Hamann's letter to Lindner of October 12, 1759).[85]

It is very difficult for Kant to transfer himself into a different philosophic idiom (philosopheme): this is very characteristic for an original thinker.

Beautiful words against the theological standpoint in the context of teleology:

"It is something very absurd to anticipate enlightenment from reason and yet to prescribe beforehand which side she must necessarily favor" (KRV II section 62).[86]

Notes

1. William Wurzer, Urs Heftrich, and Jean-Luc Nancy have seriously considered aspects of the *Critique of Judgment* in relation to Nietzsche's early philosophy. However, comparatively few scholars have explored this historical connection in detail.

2.The unconscious co-operation between the parts and the "whole" of an organism is specifically addressed by Nietzsche in "*Teleologie Seit Kant*" aka "*Zur Teleologie*," Nietzsche's early dissertation draft on natural science.

3. Although Nietzsche writes against Darwinism in numerous places (KSA 12 304, KSA 13, 303-5, 315, etc.), his comments about Darwinism are not univocal, a point that Greg Whitlock has correctly made by pointing elsewhere (KSA 7 461) to Nietzsche's claim that Darwinism is true. Nietzsche does agree with Darwin's general approach of trying to understand organic life without invoking supernatural causes.

4. Nietzsche's note about the continuity between Kant and Laplace at (BAW 3 332) probably follows from Schopenhauer's comments in WWR. Nietzsche details this connection in TSK by quoting Kant's *Universal Theory of the Heavens*, a connection that he also draws in his later lecture on Leucippus and Democritus in the eighteen seventies. Greg Whitlock. *The Pre-Platonic Philosophers: Friedrich Nietzsche* (Chicago: University of Illinois, 2001), 126.

5. Empedocles refers to strange monsters that perished in the fragments: "many foreheads without necks sprang forth, and arms wandered unattached, bereft of shoulders, and eyes strayed about alone, needing brows. . . . Limbs wandered alone." Empedocles also writes, "Many creatures were created with a face and breast on both sides; offspring of cattle with the fronts of men, and again there arose offspring of men with heads of cattle and creatures made of elements mixed in part from men, in part of female sex, furnished with hairy limbs." While this is very different from Darwin's understanding of the parts of the animals, there is common ground in the view that functionality determines survival, and the "design" is really contingent, spawned by luck. Shortly before the *Birth of Tragedy*, Nietzsche began several drafts for a projected tragedy involving Empedocles.

6. In *Truth and Lie in a Non-moral Sense* Nietzsche stresses the chance origin of the human intellect and the general overestimation of the significance of the human intellect by the human species. See chapter one, page 36, note 10. Nietzsche's definition of truth as "a mobile army of metaphors, metonyms, [and] anthropomorphisms" (TL 1) exhibits a strong continuity with his analysis in TSK and BT. The portrait of the human being as an unintentional, artistically creative subject is central to Nietzsche's view of humanity within the cosmos.

7. At another point, he refers to Kant as the "Chinaman of Königsberg," a comment that is probably meant as a joke. While Nietzsche's satire and caricatures invite unorthodox views of historical philosophers, the reasons Nietzsche describes Kant as Chinese seem cryptic and obscure. At least his reference to Kant as a *Vogelscheuche* (scarecrow) suggests someone who has no real physical body or is a willfully distorted portrait itself, a straw man.

8. Makkreel explicates the significance of Kant's suggestion that the imagination actually keeps us alive while we sleep (KU 67), a moment in the Kantian project which acknowledges the importance of the image-creating faculty for life itself. Rudolf Makkreel, *Imagination and Interpretation in Kant* (Chicago: University of Chicago

Press, 1990), 101.

9. Later Nietzsche remarks, "Kant was, with such an enthusiastic intention, the true son of his century, which before any other can be called the century of enthusiasm: as he fortunately remained also in regard to its more valuable aspects (for example in the good portion of sensism he took into his theory of knowledge). He too had been bitten by the moral tarantula of Rousseau, he too harbored in the depths of his soul the idea of that moral fanaticism." (D vor 3, 1881). In the *Genealogy of Morals*, Nietzsche also takes the opportunity to ridicule Kant's aesthetics:

> so they have offered us, from the beginning, definitions in which, as in Kant's famous definition of the beautiful, a lack of any refined first-hand experience reposes in the shape of a fat worm of error. "That is beautiful," said Kant, "which gives us pleasure without interest." Without interest! Compare with this definition one framed by a genuine "spectator" and artist—Stendhal, who once called the beautiful *une promesse de bonheur* [a promise of happiness]. At any rate he rejected and repudiated the one point about the aesthetic condition which Kant had stressed: *le desinteressement.* Who is right, Kant or Stendhal? If our aestheticians never weary of asserting in Kant's favor that, under the spell of beauty, one can even view undraped female statues "without interest," one may laugh a little at their expense. (GM III 6)

10. Nietzsche's thoughts on the *Critique of Pure Reason* are by no means univocal. Although I stress Nietzsche's debt here, it would be a mistake to overlook his harsh criticisms of the text. See (GS 354, 357), (BGE 4, 11), (GM III 25), etc.

11. See Joan Stambaugh. *The Other Nietzsche* (Albany: SUNY Press, 1993), 121-2.

12. Both Kant and Schopenhauer's concepts of the thing-in-itself are plagued by the same contradiction in that it is natural (natural, but false) to assume that the appearances bear some kind of relationship to the thing-in-itself. However, any such inference would seem to depend on a causal relationship (the thing-in-itself *causes* the appearances), yet causality is a product of the phenomenal world. Nietzsche advances this criticism in FKSP in chapter two of our investigation, "The Phenomenal Bridge."

13. Of course Nietzsche draws different consequences from Kant's delimitation of the subject to the phenomenal world:

> "Truth," according to my way of thinking, does not necessarily denote the opposite of error, but in the most fundamental cases only the posture of various errors in relation to one another. (KSA 11 598)

Nietzsche's insistence that life requires inescapable errors is important for his tragic vision of the world, and it is a view that he expresses in both the 1867 critique of Schopenhauer and 1868 critique of Kant. Unavoidable perspectival errors are a precondition for life, a limitation of the application of reason to the world. Nietzsche suggests that the world is neither able to be—nor in need of—correction through dialectics. According to this view, the fantasy of the philosophic imagination to redeem the world from errors persists only as a deep delusion.

14. Kant encountered little political trouble for his analysis of the subject content of theology, primarily because Frederick the Great did not care what philosophers wrote

about, as long as they obeyed the law. Once again power relations prove to be more fundamental than free inquiry, since Kant could not have been possible in all political states in the eighteenth century. Some of his other work, such as *Religion Within the Realm of Reason Alone*, did get him in trouble with the Prussian censors. However, for the most part Kant was able to write in relative freedom without needing to sacrifice his philosophic direction for fear of his own safety or loss of livelihood.

15. See (BT Critical Glance Backward 5) where Nietzsche writes: "a philosophy that dares to move, to demote, morality into the realm of appearance—and not merely among 'appearances' or phenomena (in the technical sense assigned to these words by idealistic philosophers), but among 'deceptions,' (*Täuschungen*) as semblance, delusion, error, interpretation, contrivance, art."

16. Wilhelm Wurzer has explored aspects of Nietzsche's mandate for intellectual integrity in "Nietzsche's Hermeneutic of Redlichkeit." *Journal of the British Society for Phenomenology*. Volume 14, Number 3 (October, 1983).

17. There is a sense of health which is biologically immanent within the physiology of the organism: the AIDS virus is not healthy for a human being. However, there is also the other sense of "health" that clearly (or sometimes not so clearly) invokes a concept that is projected via representation; the latter sense has been used to address questions about whether masturbation or prayer are healthy, and such evaluations go beyond that which is biologically immanent. The latter sense of health involved with evaluating such questions is grounded in a type of value judgment. The former sense might be thought of as existing by nature, whereas the latter sense is more clearly rooted in convention. Employing the fact-value distinction to the concept health helps clarify this issue, but the matter is further complicated by applying the concept to psychic or mental health.

18. For more on the complex problem of the physiological and its intersection with issues regarding health, see Crawford's Nietzsche's "Physiology of Ideological Criticism" In *Nietzsche as Post-modernist*. Edited by Clayton Koelb (New York: SUNY Press, 1990).

19. Nietzsche advances this view in the *Birth of Tragedy*: "Thus far we have considered the Apollonian and its opposite, the Dionysian, as artistic energies which burst forth from nature herself, without the mediation of the human artist—energies in which nature's art impulses are satisfied in the most immediate and direct way—first in the image world of dreams, whose completeness is not dependent upon the intellectual attitude or the artistic culture of any single being; and then as intoxicated reality, which likewise does not heed the single unit, but even seeks to destroy the individual and redeem him by a mystic feeling of oneness" (BT 2).

20. Eugen Fink has pursued this similarity of the Kantian sublime to the Dionysian. *Nietzsches Philosophie* (Stuttgart: W. Kohlhammer, 1960).

21. Eugen Fink, Paul de Man, and Wilhelm Wurzer have explored this possibility. However, the connection of Apollo to judgments of the beautiful and the connection of Dionysus to the feeling of the sublime has not been exhausted. Etymologically, that which appears, shines, and is beautiful is closely linked, a kinship to which Schopenhauer, Wagner, and Nietzsche were attuned. See *Paraerga and Parapelomena*. Translated by E.F.J. Payne (Oxford: Clarendon Press, 1974) for Schopenhauer's comments on the *Schöne*, *Schein*, and *Schau*, p. 211. Nietzsche also uses this etymology to develop an eye vs. ear metaphysics in *The Dionysian Worldview* (KSA 1 563), an important aesthetic constellation that Nietzsche uses in his development of his concepts of the Apollonian and Dionysian.

22. Martin Heidegger. *Nietzsche Lectures*. Volume 1 (San Francisco: Harper & Row, 1984), 111. Similar interpretations are evident in Daniel Brazeale and Gerd-Gunther Grau's commentaries on Nietzsche. Overlooking Nietzsche's work in *Teleology Since Kant*, Grau suggests that Nietzsche did not even read Kant firsthand, a claim that is indefensible. Professor Brazeale follows a similar path, but accentuates Lange as well as Schopenhauer, ruling out any "profound first hand acquaintance" of Kant by Nietzsche. Dan Brazeale, *Philosophy and Truth: Selections from Nietzsche's Notebooks of the Early 1870s* (Atlantic Highlands, N.J.: Humanities Press, 1994), 32. Even if Nietzsche frequently relied on lectures on the history of philosophy from authors such as Lange, Fischer, and Zeller, the view that Nietzsche has a superficial understanding of the history of Western philosophy due to mediation through secondary sources seems to lack evidence. It is possible that Nietzsche did not read the *Critique of Pure Reason*, but Nietzsche did read the *Critique of Judgment*.

23. At this late date it is well known that during the Third Reich Nietzsche was used for propaganda purposes by the Nazis, a use made possible in great measure by Elizabeth Nietzsche. Walter Kaufmann has shown why Nietzsche would almost certainly reject Nazism, given Nietzsche's negative comments about German culture and politics, as well as his rejection of anti-Semitism in his published works, personal life, and correspondence. It seems that various interpreters have been able to find whatever they hoped to find in Nietzsche, evidenced by Heidegger's claim that Nietzsche is offering another ontology. Nietzsche's Heraclitean claim that, "being is an empty fiction" seems to undermine Heidegger's interpretation. The Nazi propaganda machine used whatever it could to distort, and at times, rewrite history. Kaufmann's work still remains important in providing convincing evidence that Nietzsche could never be a willing accomplice to the final plan of the Nazis. Of course such vindication does not extend to Heidegger, for even after the Third Reich fell, Heidegger never repudiated his Nazism and never thought it necessary to address what significance the death chambers and lampshades made of human skin could hold for the history of being.

24. Schools of skepticism have almost never formed organizations to further any political agendas due to the general disdain of dogmatism and fanaticism, a connection of theory to practice which still has received comparatively little attention.

25. At times this question overlaps with Nietzsche's concern with the concept of health. If we were permitted to view Zarathustra as a privileged voice of Nietzsche, the mandate to remain faithful to the earth suggests that relying on another world for purpose and meaning in this life serves to slander, ignore, or negate the reality of earthly embodied existence. Ultimately Nietzsche suggests that fixation on afterworlds can be unhealthy, a type of sickness that finds expression in but is not confined to priestly castes.

26. Perhaps Nietzsche's comments on contradiction lying within the interpretation (rather than in the text or the writer himself) could be applied to Marx's interpretation of Democritus. Is the contradictory nature of Democritus's view a feature of his atomistic system, or is the contradiction merely a pseudocontradiction generated by an oversimplification and misunderstanding of Democritean nuances?

27. We refer to Homer, who very well may have been a real person who at one time told or invented these stories, but the Homeric tales first had to be written down and then they were tweaked by later compilers and editors. Thus the images we have of Homer's unified identity and style are derived representations which passed through multiple hands before arriving at their current state. They are identities derived from human

organizations, a problem that had already received considerable attention in the later ancient world.

28. Although Nietzsche usually uses the term "Intellekt" (BAW 3 374, 387, 388) or "Vernunft" in critiquing the "organ for knowledge," he also at times employs "Verstand" (BAW 3 379), (KSA 7 19 [66]), etc.

29. Schopenhauer certainly notices this dissonance and uses it to explicate and contribute to his own version of the incongruity theory of laughter, a view of laughter that is at least as old as Aristotle (Rhetoric 3 11) that appears in Kant's view as "a tense expectation transformed into nothing" in the *Critique of Judgment* (KU 54). Schopenhauer states, "laughter always signifies the sudden apprehension of an incongruity between such a conception and the real object thought under it, thus between the abstract and the concrete object of perception. The greater and more unexpected, in the apprehension of the laughter, this incongruity is, the more violent will be the laughter. Therefore in everything that incites laughter it always will be possible to show a conception and a particular, that is, a thing or event, which certainly can be subsumed under the conception, and therefore thought through it, yet in another and more predominating aspect does not belong to it at all, but is strikingly different from everything else that is thought through the conception" (Supplement to Book I: Chapter 8, "On the Theory of the Ludicrous" WWR). Schopenhauer's appraisal of laughter accentuates the gap between concepts and reals, an important theme for Nietzsche's development of the view that conceptual thought simplifies and alters the world by virtue of representation. More specifically, the gap between the real and ideal is what is at stake. The movement between the real and ideal is specifically conceived of as a feature of reflective aesthetic judgment. Both Kant and Schopenhauer use this dissonance to account for the origin of laughter, an illuminating but incomplete approach to account for this important dimension.

30. The term "aesthetic" has the dual sense of both referring to sense perception (as in the case of the Kantian transcendental aesthetic), as well as the sense of "aesthetic" that relates to judgments of the beautiful and the sublime which are usually associated with art. Nietzsche's interpretation of the activity of the transcendental subject as fundamentally form-giving in nature suggests that these two distinct senses of the term "aesthetic" actually may not be completely different. In other words, the organizational processes of the knowing subject impose form on sense data, thus revealing a view of the knowing subject that is creative, a deep aesthetic condition which makes a human perspective possible.

31. This view clearly is consistent with Nietzsche's later comment: "Parmenides said, 'One cannot think of what is not', we are at the other extreme and say 'what can be thought of must be a fiction'" (WP March-June 1888).

32. Kant does retain a moral proof for the existence of God in his system, but this is not a proof in the ordinary sense of proof. The moral proof is actually a postulate of practical reason to accommodate the view that God may be an object of rational faith. Such a proof resembles a presupposition or a premise, rather than the ordinary sense of what philosophers consider to be proofs.

33. See Schelling on the third critique in *The History of Modern Philosophy*, where he asserts that it is "Kant's deepest work, which if he could have begun with it in the way he finished with it, would have probably given his whole philosophy another direction" Translated by Andrew Bowie (Cambridge: Cambridge University Press, 1994), 173. Also, see Hegel's suggestion in *Lectures on the History of Philosophy*. Translated by R.F. Brown

and J. M. Stewart (Berkeley: University of California Press, 1990) that the *Critique of Judgment* is the "most speculative" (in the Hegelian "good" sense of speculation) of Kant's works.

34. For explication of Kant's fallout with Herder in regards to this issue, see chapter one of John Zammito's book, *The Genesis of the Critique of Judgment* (Chicago: University of Chicago Press, 1992), 33: Zammito stresses that from "the very outset, Kant conceived of aesthetic imagination as a dangerous capacity to project the unreal upon the actual."

35. Kant's earlier text, *Observations on Feeling of the Beautiful and the Sublime*, bears a resemblance to the critique of aesthetic judgment in name, but beauty and the sublime are empirical in this pre-critical text. The text is included with the version of the *Critique of Judgment* which Nietzsche used while working on his early dissertation attempt, the Rosenkranz edition (Leipzig) of 1838.

36. Nietzsche refers to Schopenhauer's account of the conditional nature of the individuating principle by citing Schopenhauer's comparison of individuation to the sailor on the violent sea: "Just as a boatman sits in his small boat, trusting his frail craft in a stormy sea that is boundless in every direction, rising and falling with the howling, mountainous waves, so in the midst of a world full of suffering and misery the individual man calmly sits, supported by and trusting the *principium individuationis*, or the way in which the individual knows things as phenomenon. The boundless world, everywhere full of suffering the infinite past, in the infinite future, is strange to him, is indeed a fiction. His vanishing person, his extensionless present, his momentary gratification, these alone have reality for him; and he does everything to maintain them, so long as his eyes are not opened by a better knowledge" (WWV I 36). The imagery of the sailor as the principle of individuation compared to the violent storm of the sea is used by Nietzsche to explicate the Dionysian. Such imagery is similar to Kant's description of the sublime in that both the sublime and Dionysian represent underlying forces of nature that make the transcendental subject feel insignificant.

37. Nietzsche's early work examines the teleological view of the universe that has been widely accepted and promoted by the long history of theological errors. According to Nietzsche's view, such errors have contributed to a false view of corporeality, organic nature, and the significance of the human species. Even if one were to prove that supernatural teleological views are bogus, such a proof is not likely to be widely received, given the nature of systematic brainwashing from the cradle onwards that many organized religious views endorse. Nietzsche suggests that the psychological programming that makes possible the belief in a god who has offspring with mortal women (as well as belief in virgin births) is something that makes those programmed decidedly hostile to the deeper truth that human beings have no special destiny in the cosmos.

38. Paley's defense of the teleological proof points to the complexity of the eye as exhibiting at least as much design as a watch found in the forest, suggesting that no one would claim a watch did not have a watch-maker. Darwin accounts quite explicitly for the existence of the eye, but his account assumes that life is much older, having far more generations than Paley could ever conceive. In spite of Paley's lack of understanding of this point, Paley and Kant still make an important observation of how common it is for organisms to maintain and heal themselves by having their internal parts work together, something that is not demonstrated within the inorganic (a clock will not grow back a broken gear or hand). This aspect of the biological organism suggests the obvious view

that earth science is not identical with biological science, even though teleological judgments have been applied historically to both domains.

39. Makkreel has pointed out that reflective and constitutive judgments need not be in opposition for Kant in "Regulative and Reflective Uses of Purposiveness in Kant." *Southern Journal of Philosophy*. Volume XXX, Supplement (1991).

40. Friedrich Lange. *Geschichte des Materialismus* (Iserlohn, Germany: Von J. Baedecker, 1866), 405-6.

41. Lange writes "All teleology has its root in the view that the builder of the universe acts in such a way that man must, on the analogy of human reason, call this action purposive." Lange, *Materialism*, 402. Lange provides some commentary on Darwinism and biology in this section of the first edition, but does not offer the extended analysis presented in the second edition of the *History of Materialism*. Nietzsche is working at this time only with the first edition, as the second edition is published in the 1870s.

42. Nietzsche's dismissal of the possibility that any unproblematic unity can ever be given is central to his theory of the understanding, not only in "*Zur Teleologie*," but throughout his corpus: "Everything that enters consciousness as 'unity' is already tremendously complex: we always only have a semblance of unity" (WP 489).

43. Nietzsche notes that Kant realizes that "nothing in the relationship of purposiveness gives it a similarity to the organism" (KU 65). Kant initially acknowledges that this way of representing organic life according to purposive causes is only a reflective postulation. It is a way of presenting the world, but such a mode of judgment is non-cognitive in nature. The world, or in this case, the organism may be completely different than how it is represented. This is why teleological judgments about "natural purposes" have a deep affinity to aesthetic judgments, a point that is important for Nietzsche's view of cognition.

44. Aristotle claims that poetry is more significant than history because history "describes the thing that has been, and poetry describes a kind of thing that might be. Hence poetry is something more philosophic and of graver import than history, since its statements are of the nature rather of universals, whereas those of history are singulars" (Poetics II 9). During Aristotle's time, very little history had been written so he may have overestimated the view that history is about singular events. After Hegel there are many modern views that suggest that aspects of history have a tendency to repeat themselves, so history too can be conceived of as philosophically important according to Aristotle's criteria. In terms of a connection between Kant and Aristotle, the sense of the poets' ability to convey a universal rather than deal only with singular events indicates a capacity to see the one in the many. This activity of the poet is made possible by an active function of the imagination that Kant explores in the context of reflective judgment.

45. This theme appears as perhaps the most fundamental question for the Socratic-Platonic dialectic in the *Birth of Tragedy*, i.e., what are the limits for conceptual thought? In this respect, Nietzsche can be interpreted as furthering the Kantian project of exploring the limits of human reason.

46. Kant's claim that "Thoughts without content are empty" (KRV A51) anticipates Nietzsche's view. Leibniz's distinction between truths of fact vs. truths of reason is important to both Hume and Nietzsche's critiques of human knowledge. Truths of reason are without content, referring only to formal reasoning operations, or relationships between pure ideas, so they are not problematic for Nietzsche. Rather, the problem lies

precisely in applying the formal truths of pure logic and mathematics to the physical world of Heraclitean flux.

47. Nietzsche's stress that there are no opposites, "things," or "x's" apart from that which we construct thus makes room for chance, as we are always located within the unknown. Much later in the *Will to Power* notes Nietzsche advances a similar argument against the noumenal that focuses on the thing-in-itself: "If I remove all the relationships, all the properties of a thing, the thing does not remain over, because thingness has only been invented by us" (WP 558).

48. Reid's example of a ship that gradually has every part of it replaced but is still referred to by the same name overlaps with Nietzsche's critique of identity, but Reid refers to the parts that make up the identity of the ship (mast, sails, etc.) in terms of visible macroproperties. Nietzsche presupposes Reid's view, but he also extends this critique of identity into the microproperties (features of cells and atoms) which may exhibit regularities, but still do not offer any fixed identities independent from the representation. See Nietzsche's November 14, 1881 letter to Overbeck for a later variation on this theme: Nietzsche compares their lives to boats afloat on the sea, temporary structures which are eventually fated to capsize.

49. This theme is reiterated in *the Birth of Tragedy* in Nietzsche's comments about God's absolute truth relegating art to the realm of illusion and falsity.

50. See Nietzsche's letter to Deussen written in the end of April or the beginning of May 1868 (BW 568).

51. See Nietzsche's May 3, 1868 letter to Rohde (BW 569).

52. Nietzsche mentions Kohl's philosophy dissertation on Kant (Immanuel Kant's *View of the Freedom of the Will*) to Rohde in an April 3, 1868 letter (BAW 565), shortly before Nietzsche decides on his own dissertation theme on Kant. It is extremely likely that Nietzsche would have mentioned that he (Nietzsche) too was working on a Kant dissertation when writing Rohde on April 3, if in fact he had begun TSK that early. For this reason it is very probable that Nietzsche began TSK sometime in April 1868. It is virtually certain that Nietzsche abandoned TSK as a formal project by September 19, 1868 (if not earlier), as indicated in his letter to Ritschl (BAW 589) in which Nietzsche suggests several other possible dissertation themes. In a later October 8, 1868 letter to Rohde (BAW 591), Nietzsche writes that he has a Homer and Hesiod dissertation planned, so it is extremely probable that Nietzsche had abandoned TSK as a formal dissertation project prior to meeting Wagner on November 8, 1868. Nietzsche was awarded his doctorate on March 23, 1869 in the following year without writing a dissertation, based on the merit of his prior publications.

53. Although Nietzsche may have been investigating teleology earlier, there is no evidence that Nietzsche was working on *Teleology Since Kant* in 1865, as Whitlock claims. Whitlock. *Pre-Platonics*, 184.

54. *Kritik der Urtheilskraft* (Hamburg: Meiner, 1990). (KU 65). All references are by section number.

55. This essay has been translated in *Friedrich Nietzsche on Rhetoric and Language* (New York: Oxford University Press, 1990), but Nietzsche's citation of the *Critique of Judgment* is misrendered (211) as the *Critique of Pure Reason*.

56. See Kant's second introduction, section 2.

57. Already during this early phase Nietzsche has attempted to dispense with the idea of an absolute object that exists independently from the relationship of a

representing being. Hence his rejection of the thing-in-itself, and the post-Boscovichian viewing of the atom as a center of force that has no absolute existence.

58. Beyond Democritus, there is also significant evidence that Nietzsche's thought of the eternal return may be indebted to the ancient atomism of Lucretius, a connection which George Stack and others have explored.

59. The other English translation of *Teleology Since Kant* is an earlier version of the one appearing here, published as *Nietzscheana* #8 (Urbana: North American Nietzsche Society, 2000).

60. See "Nietzsche on Teleology and the Concept of the Organic." In *International Studies in Philosophy*. Volume XXXI, Number 3 (1999). Professor Crawford misrenders "casualism" as "causality." In opposition to nonmaterial teleological explanations, Nietzsche claims casualism and mechanism may explain the existence of organic life. Historically casualism was attributed to the ancient atomists, a connection illustrating Nietzsche's application of the methods and views of Democritus to the problem of organic life. Crawford's study is valuable in its recognition that Nietzsche's enigmatic primordial one (*Ur-eine*) is neither primordial nor one. However, there does not appear to be any evidence to support her claim that Nietzsche was reading Hartmann's *Philosophy of the Unconscious* in 1868, as she claims on page 48 of *The Beginnings of Nietzsche's Theory of Language* (New York: de Gruyter, 1988). Nietzsche did read it in the following year, 1869, as indicated by his correspondence.

61. (WWR II [Appendix] 533).

62. Schopenhauer writes this in chapter 26 as well as in the appendix of the criticism of the Kantian philosophy in WWR. Nietzsche uses the term *unbezwinglich* rather than Schopenhauer's *unwiderleglich*.

63. Here Nietzsche is attempting to eliminate the "God's eye view," to which organic life would find correspondence with God's foreknowledge and intelligent design.

64. See Aristotle's *Physics* (2.2, 2.8) and *On the Parts of the Animals* (1.1) for what Nietzsche refers to elsewhere as the "Empedoclean insight." Empedocles held that most of the parts of animals came to be by chance. Nietzsche notes Lange's observation that "the Empedoclean insight" which Democritus assimilated corresponds to Darwin's theory (BAW 4 54). Thus Nietzsche also stresses that organisms are not the result of design, but products of chance.

65. (WWR I 157).

66. (WWR I 157). Nietzsche is rejecting Schopenhauer's claim that the will is the thing-in-itself here. Nietzsche suggests that the will that powers the living organism is not singular but a multiplicity, and what Schopenhauer refers to as the will is actually contingent on more fundamental power relations, rooted in the conditions for survival.

67. (WWR I 160).

68. Friedrich Lange. *Materialismus*. 197, 198.

69. In the Beck edition, the following passage appears which has been omitted from the Musarion: "[There is] much that is comic in the *Minor Writings* of Brockes s. Strauss in the Stoics *v.* Zeller B.4."

70. This sentence, which is not present in the Musarion edition, but appears in *Frühe Schriften* and *Werke*, prompts one to consider it in light of Nietzsche's 1869 essay, "On the Origin of Language," where he writes that "the right knowledge [concerning the origin of language] has become common only since Kant, who in the *Critique of Judgment* both recognized teleology in nature and at the same time stressed the remarkable antinomy that

something can be purposive without a consciousness." According to Proclus, "Pythagoras and Epicurus agree with Cratylus, but Democritus and Aristotle agree with Hermogenes, the former that names arise by nature, the latter that they arise by chance . . . . Democritus thought that the proof of their chance origin was fourfold: (1) the calling of different things by the same name; (2) having several names for the same thing; (3) change of name; (4) lack of name." Mary Ann Freeman, *Ancilla to the Pre-Socratic Philosophers* (Cambridge: Harvard University Press, 1948), 97. "Convenience" might better be understood as "convention" in this context, insofar as there is a chance origin of language and teleology in that both indicate something that does not exist by nature, but is a product of human convention. In any case, human language is viewed here, like organisms are, as a contingent structure.

71. The page numbers which Nietzsche cites from Schopenhauer, Goethe, and Kant have been omitted in the Musarion edition; appearing in the Beck edition and in *Werke*, the pagination cited by Nietzsche corresponds to the pagination in the Rosenkranz edition of Kant (volume 4) which also includes *Beobachtungen über das Gefühl des Schonen und Erhabenen* (Leipzig, 1838); the page numbers do not correspond to Kuno Fischer's quotes of Kant, so it appears that Nietzsche was working with Kant's third critique with a firsthand acquaintance.

72. Lange points to this same passage from Goethe in the *History of Materialism*, volume 2, section IV, "*Darwinismus und Teleologie*," emphasizing its importance for understanding organic life (2 694).

73. The parenthetical citation "for philosophy in general, intuition of the power of judgment" is only present in the Musarion edition.

74. The *Nachbericht* in the Beck edition refers this quote and the ones which follow it to Kuno Fischer's second volume on Kant in the *Geschichte der neuern Philosophie*. These passages (which are not enclosed in quotation marks in the Musarion edition) do not follow verbatim from Kuno Fischer, but appear to follow Fischer's reading closely nonetheless; i.e., these are Nietzsche's interpolations of Fischer.

75. "Organisms" are understood here as beings that have been organized by a designer.

76. In the Beck edition the following outline appears immediately after this paragraph:

I. The organism as a product of our organization

II. The mathematical alone is knowable

III. (Blank: left unfinished by Nietzsche)

77. "The casualistic principle refers matter to a physical basis for its form" (*Critique of Judgment* 72 391). Kant writes, "attributed to Epicurus or Democritus—it is so manifestly absurd, if taken literally, that we must not let it detain us."

78. This sentence does not appear in the Musarion edition.

79. This page citation refers to the quote which immediately follows from the fourth volume of the Rosenkranz edition of Kant. This page number appears neither in Schopenhauer nor in Fischer's analyses of Kant, so it is not as if Nietzsche was simply copying a secondary source. Nietzsche must have read Kant's *Critique of Judgment* directly, since his pagination citation corresponds to the 1838 Rosenkranz edition.

80. The following outline from the Musarion edition is here omitted from the Beck version, but the latter affixes it to the end of the draft:

Chapter 1. Concept of purposiveness (as capacity for life).

Chapter 2. Organism (the indeterminate concept of life, the indeterminate

	concept of the individual).
Chapter 3.	The ostensible impossibility of mechanistic explanation of the organism (what does mechanistic mean?)
Chapter 4.	The recognized purposelessness of nature in contradiction to purposiveness.

81. The analogy of the relationship between letters and a tragedy to the parts of the animal and the whole organism also finds expression in Richard Dawkins' work to explain cumulative selection. In *The Blind Watchmaker* (New York: W.W. Norton, 1986), Dawkins responds to contemporary intelligent design theorists by showing how the astonishing complexity of the hemoglobin molecule can be explained without invoking a conscious designer. Nietzsche's reflections on the parts of the animals and the parts of language also attempt to show that in both cases the complex identities that are created by the parts (whether organic or literary) are actually contingent in nature.

82. This may refer to Kant's critique of Spinoza's concept of transcendental perfection of a thing, which Kant calls a "childish play with words rather than concepts" (KU 73).

83. The Musarion edition has the following outline here which the Beck version affixes to the end of the draft:

Chapter 1:	Teleological observation is observation according to forms.
Chapter 2:	Forms (individuals) belong to and are inferred from the human organization.
Chapter 3	Lifeforce.

84. Compare with *Philosophy in the Tragic Age of the Greeks*, sec. 17 and Nietzsche's lecture on Leucippus and Democritus in Whitlock. *Pre-Platonics*, 126. This quote appears in Kuno Fischer's reading of Kant in *Geschichte der Neuern Philosophie*, Volume 5, (Heidelberg: Carl Winter, 1899) 494.

85. Fischer, *Geschichte*, volume 4, 211.

86. This passage also is quoted by Schopenhauer in *On the Will in Nature*. After several outlines (see endnotes 79 and 82) at the end of TSK, Nietzsche lists a bibliographical compilation of works to be used in his natural science dissertation:

Kant, *Allgemeine Naturgeschichte und Theorie des Himmels* 1755.
Der einzig mögliche Beweisgrund zu einer Demnostration des Daseins Gottes.
Holbach *Système de la nature.*
Treviranus *[Über] die Erscheinungen und Gesetze des organischen Lebens* 1832
Czolbe *neue Darstellung des Sensualismus* Leipz. 1855
Die Grenzen und der Ursprung der menschl. Erkenntniß Jena und Leipz 1865.
Moleschott *Kreislauf des Lebens* 1862.
Die Einheit des Lebens Giessen 1864.
Virchow *4 Reden über Leben und Kranksein* Berlin 1862.
gesamm Abhandl. Zur wissen. Med. *Frankf.* 1856.
Trendelenburg *Logische Untersuchungen* Leipz. 1862.
Überweg *System der Logik*
Helmholtz *überdie Erhaltung der Kraft* Berlin 1847.
Über die Wechselwirkung der Naturkräft 1854.
Wundt *Vorlesungen über die Menschen- und Thierseele.*

Lotze *Streitschriften* Leipzig 1857
Medicin. Psychologie 1852
Trendelenburg *Monatsber. Der Berl. Acad.* Nov. 1854
Febr. 1856.
Historisch Beiträge zur Philosophie 1855
Herbart *analyt. Beleuchtung des Naturrechts und der Moral.*
Schelling *Ideen zu einer Philosophie der Natur*
Herder *Ideen zur Philos. der Gesch. Der Menschheit.*
Bichat *sur la vie et la mort* (*zu lesen*).
Joh. Müller *über das organ. Leben*
über die Physiologie der Sinne.

Kant *Kritik der Urtheilskraft* 1790.
Fries *mathem. Naturphilos.* Heidelberg 1822
Schleiden *über den Materialis. in der neueren Naturwissenschaft.* Leipzig 1863.
(bei Schleiden mechan. Erklärbark. Der Organismen)
C. Rosenkranz, *Schelling Vorles.* Danzig 1843.
Sal. Maimon 1790 (*Berl. Journal f. Aufklärung* von A. Reim Bd. VIII St. 1)
Schelling *System des transcendent. Idealismus.*
Oken *die Zeugung* 1805
Lehrb. *Der Naturphilosophie* 1809 II Aufl. 1843
Carus *Grundzüge der vergl. Anatomie und Physiologie* 1825.

Conclusion: Aesthetic of Becoming

Nietzsche's philosophical genealogy remains a peculiar one. His broad interests in aesthetics, philology, psychology, and natural science mark him as a philosopher who was influenced by many diverse thinkers, many of whom are typically not considered to be "philosophers." Nonetheless, Nietzsche also dialogues with key figures within the Western philosophical tradition that allowed him to carve out his own niche. Within the history of ideas Nietzsche may have had the broadest influence of any writer in the West, given his influence within philosophy and beyond. Few other philosophers have exerted significant influence in literary theory, psychology, etc.

This inquiry has explored some of the important philosophical thinkers that have contributed to the shaping of Nietzsche in his development as a young philosopher—hence the title *Becoming Nietzsche*. These pivotal thinkers helped Nietzsche to become what he was. Without marginalizing the creativity and liveliness from his later polished writings, we may still view Nietzsche as a synthesis of different streams of diverse thought. These influences helped create the content of his enigmatic philosophical reflections. Even if tracking down Nietzsche's early mentors can never adequately clarify Nietzsche's creative synthesis, such investigation does elucidate how some of Nietzsche's themes for investigation are part of a continuing historical philosophic dialogue. Moreover, these early reflections indicate that Nietzsche had already developed a rigor in his philosophic methods by 1867, and they help to shed light on how original a thinker Nietzsche is.

Nietzsche's early philosophic encounters with Democritus, Schopenhauer, and Kant help explain the conditions that lead to his peculiar philosophical views, a complex synthesis of modern and ancient views. Although such early juvenilia can never be the final word on Nietzsche, these early reflections illuminate his development as a serious thinker in the 1860s. This early work indicates the extent to which Nietzsche was concerned with the significance of human reason within the cosmos. Nietzsche's focus on this triad of thinkers from 1866 to 1868 indicate the deep influence that Friedrich Lange exerted on Nietzsche's genealogy before Nietzsche would meet Richard Wagner.

In spite of the incomplete, unpolished, and fragmentary nature of his work from these late 1860s philosophic encounters, there still is something present in these pages that helps us understand how Nietzsche developed as a philosopher, thus contributing to how we are to understand what the term

"Nietzschean" means. The word "Nietzschean" continues to be ambiguous, given that Nietzsche has been associated with music, nihilism, amoralism, the Antichrist, the death of God, aesthetic excess, will to power, etc. There probably have been more diverse and conflicting interpretations of Nietzsche than any other thinker in the history of the world, as is evidenced by the very different senses of what Nietzsche's primary significance is. This is not to suggest that all interpretations of Nietzsche are equally plausible. Rather, this work hopes to show that the view of Nietzsche as a philosopher of becoming (in the Heraclitean sense) is justified, in spite of the fact that Nietzsche is only occasionally thought of in that sense.

Above all other thinkers in the Western tradition before him (excluding his contemporary Lange) Democritus is one of Nietzsche's few genuine mentors. Even if the ancient image of Democritus as the laughing philosopher is of a later origin, Democritus still shares Zarathustra's crowning virtue.[1] In late 1867 or early 1868 Nietzsche writes:

> Where does the story come from that Democritus used to laugh at all things? Someone had thought Democritus and Heraclitus were in opposition! (BAW 3 333)

How could it be thought that Democritus and Heraclitus were in opposition? Perhaps it was the story of Democritus as the laughing philosopher that was put in opposition to the *character* of Heraclitus as the weeping philosopher, a juxtaposition that seems foreign to Nietzsche's view. Both Democritus and Heraclitus are mentors outside of the history of Platonism for Nietzsche, each providing a view that is presupposed in Nietzsche's interpretation of a new meaning for the earth. Each provide a view that is free from a certain type of Platonic idealism and ontotheology. Nietzsche never rejects Heraclitus's view of change, unlike his critical encounters with Kant and Schopenhauer in the 1860s.

Heraclitus's view of the world as flux resonates deeply with Nietzsche's aesthetic view of the world, a view Nietzsche puts forth in TSK in 1868. Nietzsche's commentary on Heraclitus in his pre-Platonic lectures several years later explicitly reveals the deep Heraclitean-Nietzschean affinity that is suggested in the dissertation draft on Kant:

> Nature is just as infinite inwardly as it is outwardly: we have succeeded up to the cell and to parts of the cell, yet there are no limits where we could say here is the last divisible point. Becoming never ceases at the indefinitely small. . . Well, this is the intuitive perception of Heraclitus; there is no thing of which we may say, "it is." He rejects Being. He knows only Becoming, the flowing. He considers belief in something persistent as error and foolishness.[2]

The view that change is more fundamental than same, that becoming and transformation are more fundamental than any literary, philosophic, or scientific concepts of identity, remains as an important part of Nietzsche's tragic vision of the world. In his thinking about Heraclitus Nietzsche finds himself, claiming that

> the Heraclitean eternally living fire may be grasped only as aesthetic phenomenon. We find here a purely aesthetic view of the world. We must exclude even more any moralistic tendencies to think teleologically here, for the cosmic child (Weltkind) behaves with no regard to purposes but rather only according to an immanent justice: it can act . . . lawfully, but it does not will these ways. That constitutes the abyss between Heraclitus and Anaxagoras, and that is the point that more recent commentators have failed to understand.[3]

Nietzsche pays tribute to Heraclitus in his attempt to restore the innocence of becoming by advancing his own attempt to view the world only as aesthetic phenomenon. If Nietzsche's view of the tragic nature of the world is correct, the cosmos is free from any supernatural punishment and reward systems that calculate the guilt of transgressors to be exacted in future incarnations as part of a moral world order. Nietzsche already begins to articulate this worldview in the 1860s in his notes on Democritus. Such a view would later find expression in Zarathustra's injunction to remain faithful to the earth.

Nietzsche rejects foundational atomistic materialism, but he never rejected the philosophical mentor who claimed "We know nothing in reality, for truth lies in an abyss." The deep skepticism built into Democritus's theory of perception need not imply a self-refuting paradigm, but may suggest a poetic view of the world that acknowledges how very little humans know. As one of the greatest masters of the aphoristic style, Democritus was able to write poetry and philosophy simultaneously as a hybrid art. Democritus inspired Nietzsche like no other thinker could.

Democritus's writings on friendship, pleasure, mathematics, style, poetry, and ethics fascinated Nietzsche, providing him with a view of health and simplicity that is not at odds with intellectual rigor and achievement. Nietzsche's comparison of Democritus to Aristotle's ethics stresses a type of reverse fanaticism in character development, a kind of character formation that does not flee to the supernatural for concepts of meaning and virtue within the human world.

In Nietzsche's early writings on Democritus he begins to suggest a theory of revenge as an underlying motive for Platonism and the religious metaphysics that postulate a moral-world order. The tragic play of becoming in which all organic life finds itself is embraced and affirmed by Nietzsche, a love of fate that is not resignation, but affirmation. This stress on life-affirmation distinguishes Nietzsche in his later years from Schopenhauer, since Nietzsche poses this as the challenge that breaks from Schopenhauer's view of the world as unworthy of affection. How is one to affirm life, even in its most horrific and terrifying moments? Nietzsche provides a hint with music, laughter, and dancing, the arts of this-worldly affirmation.

Although Nietzsche focuses on Kant's "Critique of Teleological Judgment" in his 1868 analysis of Kant largely to show how Kant is not entitled to his expanded presentation of teleological order, Nietzsche still recognizes that Kant had made teleological judgments problematic in an important sense in the history of philosophy. Their status as reflective judgments within the Kantian critical enterprise revokes any direct claim to cognition that such claims could make. Kant's reflection on organic life raises fundamental questions about how organic beings represent embodiment, an important theme that animates Schopenhauer's philosophy in its reflections on discursive epistemological limits.

Earlier in the *Critique of Judgment* (before the "Critique of Teleological Judgment" in the "Critique of Aesthetic Judgment"), Kant offers additional observations on embodiment that suggest a similarity to Nietzsche's own thoughts on laughter. After Kant's description of laughter as "an affect that arises if a tense expectation is transformed into nothing" (KU 54), Kant criticizes Voltaire for not mentioning laughter as something important to counterbalance the many hardships of life. Kant writes:

> [M]usic and something to laugh about are two kinds of play with aesthetic ideas, or for that matter with presentations of the understanding by which in the end nothing is thought; it is merely the change they involve that still enables them to gratify us in a lively way. This shows rather clearly that in both of them the quickening is merely bodily, even though is it aroused by ideas of the mind, and shows that all gratification [we find] at a lively party, extolled as being so refined and inspired consists [merely] in the feeling of health that is produced by an intestinal agitation corresponding to such play. (KU 54)

In Kant's later years he advances laughter as medicine insofar as it produces the feeling of health in the body. Although this might seem to be a strange view of Kant, this is not an aberration in Kant's thought, since he proclaims laughter even more vigorously in his post-critical work, the *Anthropology*:

> It does not matter who makes us laugh—a hired jester (harlequin) or an artful scamp among our circle of friends, "a sly dog" who seems to have no mischief on his mind and does not join in the laughter, but with seeming simplicity suddenly releases our strained anticipation (like a taut string). [Whatever provokes it,] laughter is always a shaking of the muscles involved in digestion, which promotes it far better than the physician's wisdom would do. (Anth 79)

As something greater than a physician's wisdom, the view that laughter can restore the health of the body reveals a side of Kant that seems much more akin to Nietzsche's view of the art of this-worldly comfort.[4] Even if Kant offers us some similar aspirations to Nietzsche's idea of the laughing sage, Kant and Nietzsche do have different views about the role and nature of the human mind.[5] Although much of Kant's philosophy seems discordant with Nietzsche's philosophic interests, the *Critique of Judgment* exhibits some similarities to the mood (*Stimmung*) of Nietzsche's work nonetheless, as indicated in chapter three of our inquiry. Kant's portrait of aesthetic judgment as grounded in the feeling of life (*Lebensgefühl*) suggests a close proximity between art and life, a kinship which is explored further by Nietzsche's aesthetic perspectivism and desire to offer a philosophy in the service of life.

This investigation into Nietzsche's philosophic encounters with Democritus, Schopenhauer, and Kant shows that Nietzsche enters into dialogue with important philosophers within the Western tradition in a serious and meaningful way in the late 1860s. These three thinkers helped shape Nietzsche's philosophic orientation, contributing to how he understood and evaluated the significance of human reason, tragedy, and organic life. Although this early phase in the genealogy of Nietzsche's philosophy makes possible a more complete and rigorous understanding of how he developed as a philosopher, it has received little attention. Nonetheless, this early phase of Nietzsche's philosophy is important, since it must be considered to assess accurately how original Nietzsche is as a thinker.

Notes

1. The life-affirming spirit of the laughing sage is shared by both Democritus and Nietzsche's Zarathustra: "This crown of laughter, the rose-wreath crown: I crown myself with this crown; I myself pronounced holy my laughter. I did not find anyone else today strong enough for that. This crown of laughter, the rose-wreath crown: to you, my brothers, I throw this crown. Laughter I have pronounced holy: you higher men, *learn*—to laugh" (Z IV).

2. Greg Whitlock. *The Pre-Platonic Philosophers: Friedrich Nietzsche* (Chicago: University of Illinois Press, 2001), 62.

3. Whitlock. *Pre-Platonics*, 70.

4. Kant's observations about the limits of the physician's wisdom to assist in digestion could be further dialogued with Nietzsche's own reflections on the tension between the Dionysian and the Apollonian. Kant specifically connects the shaking of the muscles in digestion to bodily health, an apparent way to access the feeling of bodily health without Apollonian mediation.

5. For that matter, the significance of laughter for these two thinkers occupies different places in their thought. Moreover, there are of course multiple types of laughter, not all of which would indicate a type of life-affirmation for Nietzsche.

Selected Bibliography

Allison, David. "Nietzsche Knows No Noumenon." In *Why Nietzsche Now?* Bloomington: Indiana University Press, 1985.

———. *Reading the New Nietzsche*. Oxford: Rowman & Littlefield, 2001.

Ameriks, Karl. "Recent Work on Kant's Theoretical Philosophy" In *American Philosophical Quarterly*. Volume 19, Number 1, January 1982.

Ariew, Andrew, Cummins, Robert and Perlman, Mark. *Functions: New Essays in the Philosophy of Psychology and Biology*. New York: Oxford University Press, 2002.

Aristotle. *The Basic Works of Aristotle*. New York: Random House, 1941.

Behe, Michael. *Darwin's Black Box*. New York: Free Press, 1996.

Böning, Thomas. *Metaphysik, Kunst, und Sprache beim Frühe Nietzsche*. New York: Walter de Gruyter, 1988.

Borchmeyer, Dieter. *Theory and Theatre*. Oxford: Oxford University Press, 1991.

Brazeale, Daniel. *Philosophy and Truth: Selections from Nietzsche's Notebooks of the Early 1870s*, edited and translated by Daniel Brazeale. Atlantic Highlands, New Jersey: Humanities Press. 1994.

Brown, Richard. "Nietzsche and Kant On Permanence." In *Man and World*. Volume 13, 1980.

Bueb, Bernhard. *Nietzsches Kritik der Praktishen Vernunft*. Stuttgart: Ernst Klett, 1970.

Clark, Maudemarie. *Nietzsche on Truth and Philosophy*. Cambridge: Cambridge University Press, 1990.

Copernicus, Nicholas. "On the Revolution of the Heavenly Bodies." In *The Philosophy of the 16th and 17th Centuries*, edited by Richard H. Popkin. New York: Free Press, 1966.

Crawford, Claudia. *The Beginnings of Nietzsche's Theory of Language*. New York: Walter de Gruyter, 1988.

———. "Nietzsche's Physiology of Ideological Criticism." In *Nietzsche as Post-Modernist*, edited by Clayton Koelb. New York: SUNY Press, 1990.

Dannhauser, W.J. *Nietzsche's View of Socrates*. Ithaca, New York: Cornell University Press, 1974.

Deleuze, Gilles. *Kant's Critical Philosophy*, translated by Hugh Tomlinson and Barbara Habberjam. Minneapolis: University of Minnesota, 1993.

———. *Nietzsche et la Philosophie*. Paris: Presses Universitaires de France, 1956.

Descartes, René. *Selected Philosophical Writings*. Translated by Norman Kemp Smith. New York: Random House, 1958.

Desmond, William. "Schopenhauer, Art, and the Dark Origin" in *Schopenhauer: New Essays in Honor of his Two Hundredth Birthday*. Lewiston, New York: Mellon Press, 1988.

Dickopp, Karl-Heinz. "Aspekte Zum Verhaeltnis Nietzsche-Kant und Ihre Bedeutung für Die Interpretation des 'Willen Zur Macht.'" In *Kant Studien*. 61, 1970.

Diogenes Laertius. *Lives of the Eminent Philosophers*, translated by R.D. Hicks. Cambridge: Harvard University Press, 1995.

Durant, Will. *The Story of Philosophy*. New York: Washington Square Press, 1961.

Fink, Eugen. *Nietzsches Philosophie*. Stuttgart: W. Kohlhammer, 1960.

Fischer, Kuno. *Geschichte der neuern Philosophie*, vols. 4 & 5. Heidelberg: Carl Winter, 1899.

Förster-Nietzsche, Elizabeth. *The Young Nietzsche*, translated by Anthony Ludovici. London: William Heinemann, 1912.

Freeman, Mary Ann. *Ancilla to the Pre-Socratic Philosophers*. Boston: Harvard University Press, 1948.

Grau, Gerd Gunther. *Ideologie und Wille zur Macht: Zeitgemässe Betrachtungen über Nietzsche*. New York: Walter de Gruyter, 1984.

Hamacher, Werner. "The Promise of Interpretation: Reflections on the Hermeneutical Imperative in Kant and Nietzsche." In *Looking after Nietzsche*, translated by Jane O. Newman and John H. Smith. Albany, New York: SUNY Press, 1990.

Heftrich, Urs. "Nietzsches Auseindersetzung Mit Der 'Kritik der Aesthetischen Urteilskraft.'" *Nietzsche Studien*. Volume 20, 1992.

Hegel, George William Friedrich. *Lectures on the History of Philosophy*, translated by R.F. Brown and J.M. Stewart. Berkeley: University of California Press, 1990.

Heidegger, Martin. *Nietzsche*, translated by David Farrell Krell. San Francisco: Harper & Row, 1984.

Janz, Kurt Paul. "Die Kompositionen Friedrich Nietzsches." *Nietzsche Studien*. Volume 1, 1972.

Kant. Immanuel. *Kritik der Urteilskraft*. Hamburg: Meiner, 1990.

———. *Anthropology*, translated by Mary J. Gregor. The Hague: Nijhoff, 1974.

———. *Critique of Judgment*, translated by Werner Pluhar. Indianapolis: Hackett, 1987.

———. *Critique of Practical Reason*, translated by Lewis White Beck. New York: Macmillan, 1989.

———. *Critique of Pure Reason*, translated by Norman Kemp Smith. New York: St. Martin's Press, 1965.

———. *Grounding for the Metaphysics of Morals*, translated by James Ellington. Indianapolis: Hackett, 1993.

———. *Prolegomena for Any Future Metaphysics*, translated by Lewis White Beck. Indianapolis: Bobbs-Merrill, 1950.

Kaufmann, Walter. *Nietzsche: Philosopher, Psychologist, Antichrist*. Princeton. N.J.: Princeton University Press, 1974.

Kerkhoff, M. "Nietzsche Kantkritik Und Ihre Voraussetzungen." *Proceedings of the Third International Kant Congress*, 1972.

Kulenlampff, Jens. "*Materalien zu Kants 'Kritik Der Urteilskraft.'*" Frankfurt: Suhrkamp, 1974.

Lange, Friedrich. *Geschichte des Materialismus und Kritik Seiner Bedeutung in der Gegenwart*. Iserlohn, Germany: Von J. Baedeker, 1866.

———. *The History of Materialism*, translated by Ernest Thomas. Boston: Houghton, Osgood & Co., 1880.

Love, Friedrich. *Young Nietzsche and the Wagnerian Experience*. New York: AMS Press, 1966.

Lucretius. *On the Nature of the Universe*, translated by James H. Mantinband. New York: Frederick Ungar Publishing, 1972.

Lyotard, Jean-Francois. "The Interest of the Sublime." *Of the Sublime: Presence in Question*. New York: SUNY Press, 1993.

———. *Lessons on the Sublime*, translated by Elizabeth Rottenberg. Stanford: Stanford University Press, 1994.

Magnus, B. "Nietzsche's Philosophy in 1888. The Will to Power and the Übermensch." In *Journal of the History of Philosophy*. Volume 24, Number 1, January 1986.

Makkreel, Rudolf. *Imagination and Interpretation in Kant*. Chicago: University of Chicago Press, 1990.

———. "Regulative and Reflective Uses of Purposiveness in Kant." *The Southern Journal of Philosophy*. Volume XXX, Supplement, 1991.

Marcuse, Herbert. *The Aesthetic Dimension*. Boston: Beacon Press, 1978.

Marx, Karl and Engels, Friedrich, *Collected Works*. Vol. I, translated by Dirk J. and Sally R. Struik, New York: International Publishers, 1975.

Meerbote, Ralph. "Reflections on Beauty." *Essays in Kant's Aesthetics*, edited by Cohen and Guyer. Chicago: University of Chicago Press, 1982.

Mejer, Jorgen. *Diogenes Laertius and His Hellenistic Background*. Wiesbaden: Steiner, 1978.

Middleton, Chris. *Selected Letters of Friedrich Nietzsche*. Chicago: University of Chicago Press, 1969.

Mosser, Kurt. "Nietzsche, Kant, and the Thing-in-itself." *International Studies in Philosophy*. Volume XXV, Number 2, 1993.

Nancy, Jean-Luc. "Nietzsche's Thesis on Teleology." In *Looking After Nietzsche*, translated by Peter Connor. Albany, N.Y.: SUNY Press, 1990.

———. "The Sublime Offering." *Of the Sublime: Presence in Question*. New York: SUNY Press, 1993.

Neiman, Susan. *Evil in Modern Thought*. Princeton, N.J.: Princeton University Press, 2002.

Nietzsche, Friedrich. *Brief-wechsel*. New York: Walter de Gruyter, 1975.

———. *Frühe Schriften*. Stuttgart: Beck Verlag, 1994.

———. *Gesammelte Werke*. Munich: Musarion Verlag, 1922.

———. *Kritische Studienausgabe*. New York: de Gruyter, 1988.

———. *Werke*. Leipzig: Kröner Verlag, 1912.

———. *Basic Writings of Nietzsche*, translated by Walter Kaufmann. New York: Modern Library, 1968.

———. *Daybreak*, translated by R.J. Hollingdale. Cambridge: Cambridge University Press, 1982.

———. *Friedrich Nietzsche on Rhetoric and Language*. Oxford: Oxford University Press, 1989.

———.*The Portable Nietzsche*, translated by Walter Kaufmann. New York: Penguin Books, 1983.

———. *The Gay Science*, translated by Walter Kaufmann. New York: Vintage Books, 1974.

———. *The Will to Power*, translated by Walter Kaufmann and R.J. Hollingdale. New York: Vintage Books, 1967.

O'Hara, Daniel. "The Prophet of Our Laughter: Or Nietzsche As—Educator?" *Why Nietzsche Now?* Bloomington: Indiana University Press, 1985.

Paley, William. *Natural Theology*. London: J. Faulder, 1809.
Pletsch, Karl. *Young Nietzsche*. New York: Free Press, 1991.
Plato. *Protagoras and Meno*, translated by Guthrie. New York: Penguin, 1986.
———. *The Republic*, translated by Cornford. New York: Oxford University Press, 1980.
———. *Timaeus, Critias, Cleitophon. Mexenus, Epistles*, translated by R. G. Bury. Cambridge: Harvard University Press, 1999.
———. *The Laws*, translated by Trevor Saunders. New York: Penguin, 1975.
Porter, James. *Nietzsche and the Philology of the Future*. Stanford: Stanford University Press, 2000.
———. *The Invention of Dionysus*. Stanford: Stanford University Press, 2000.
Raphael, Robert. *Richard Wagner*. New York: Twayne Publishers, 1969.
Reboul, Olivier. *Nietzsche critique de Kant*. Paris: Presses Universitaires de France, 1974.
Sallis, John. *Crossings: Nietzsche and the Space of Tragedy*. Chicago: University of Chicago Press, 1991.
Sander, Hans-Dietrich. *Marxistiche Ideologie und Allgemeine Kunsttheorie*. Tübingen: Mohr, 1970.
Schelling, Friedrich *Lectures on the History of Modern Philosophy*, translated by Andrew Bowie. Cambridge: Cambridge University Press, 1994.
Schlechta, Karl. *Nietzsche-Chronik: Daten zu Leben und Werk*. München: Deutscher Taschenbuch Verlag, 1975.
Schmidt, H.J. *Nietzsche und Socrates*. Meisenheim am Glan: Anton Hain, 1969.
Schopenhauer, Arthur. *On the Will in Nature*, translated by E.F.J. Payne. New York: St. Martin's Press, 1991.
———. *Paraerga and Parapelomena*, translated by E.F.J. Payne. Oxford: Clarendon Press, 1974.
———. *The World and Will as Representation*. Translated by E.F.J. Payne. Indian Hills, Colorado: Falcon's Wing Press, 1958.
Schwartz, Steve. *What Is Freedom?* Kingston: Rhode Island Book Company, 2001.
Silk, M.S. and J.P. Stern, *Nietzsche on Tragedy*. New York: Cambridge University Press, 1981.
Stack, George. *Kant, Lang and Nietzsche: Critique of Knowledge*. New York: Routledge, 1991.
———. *Lange and Nietzsche*. New York: Walter de Gruyter, 1983.
Stambaugh, Joan. *The Other Nietzsche*. Albany: SUNY Press, 1993.
Stein, Jack. *Richard Wagner & The Synthesis of the Arts*. Detroit: Wayne State University Press, 1960.
Swift, Paul. "In-jestion: Intestinal Laughter in Kant and Nietzsche." *International Studies in Philosophy*. Volume XXVII, Number 1, 1995.
———. "Nietzsche On Teleology and The Organic". *International Studies in Philosophy* Volume XXXI, Number 3, 1999.
———. Introduction to "Teleology Since Kant" appearing as *Nietzscheana #8*, edited by Richard Schacht. 2000.
Taminiaux, Jacques. *Poetics, Speculation, and Judgment*, translated by Michael Gendre. Albany: SUNY Press, 1993.
Voltaire. *Candide, ou L'optimisme*. Paris: Didier, 1957.
Wagner, Richard. *Correspondence of Wagner and Liszt*, translated by Francis Hueffer. New York: Haskell House, 1969.
———. *Richard Wagner's Prose Works*, translated by William Ashton Ellis. London: Routledge & Kegan Paul, 1893.

Index

About the Author

Paul A. Swift received his Ph.D. from Duquesne University and teaches philosophy at Bryant University. He has published numerous articles on philosophy and is accomplished as a singer-songwriter. Dr. Swift is a certified black belt in Uechi-Ryu Karate-do by the Okinawa Karatedo Federation in Okinawa, Japan. He lives in Putnam, Connecticut.

Manufactured by Amazon.ca
Bolton, ON

24509078R00081